Exploring The BUILDING BLOCKS of SCIENCE

Book 4

LABORATORY NOTEBOOK

REBECCA W. KELLER, PhD

Illustrations: Janet Moneymaker

Exploring the Building Blocks of Science Book 4 Laboratory Notebook
ISBN 978-1-941181-06-5

Published by Gravitas Publications Inc.
www.realscience4kids.com
www.gravitaspublications.com

Contents

Experiment 1

Get Inside a Cell Phone!

Introduction

Find out what's inside your cell phone!

I. Think About It

❶ What materials do you think a cell phone is made of?

❷ How do you think science was used to make and form the materials in cell phones?

❸ What do you think scientists and engineers need to know to be able to create cell phones?

❸ Which different fields of science do you think were needed to develop cell phones? Why?

II. Observe It

1. Find an old cell phone that you can examine.
2. Carefully pull off the back cover. Examine the different components (parts). Wear rubber gloves while taking apart the cell phone.
3. Draw the inside of the cell phone.

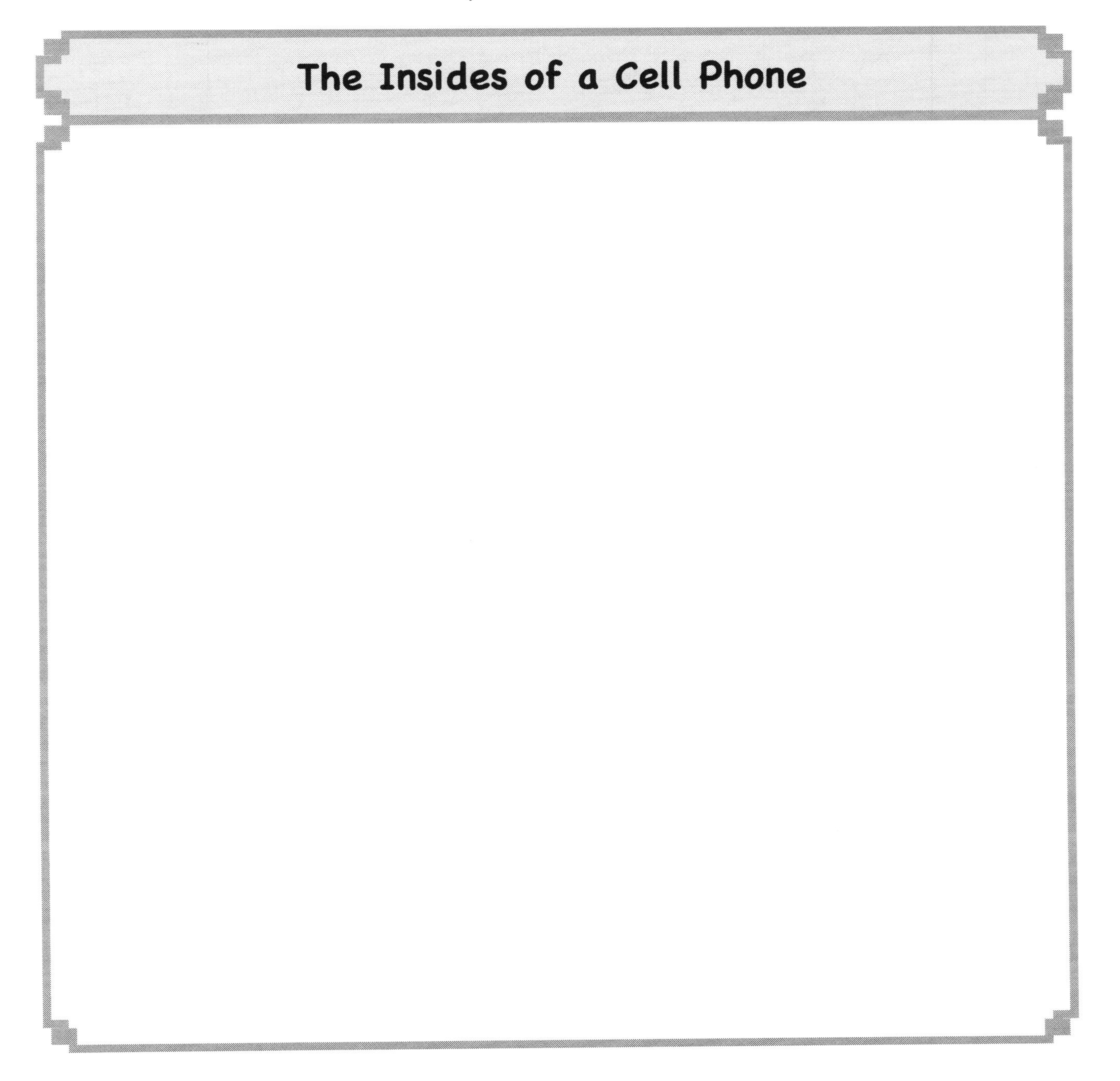

❹ Find and remove the SIM card. It should be a small, flat rectangular part. SIM stands for Subscriber Identity Module. SIM cards store data like phone numbers and passwords.

❺ Find and remove the battery. Do not open the battery.

❻ Try to carefully open the cell phone further. This will largely depend on the age and type of cell phone you are taking apart. Disassemble the cell phone as much as possible without destroying any of the parts (a hammer is not a suitable tool!).

❼ Draw all the parts of the cell phone you were able to remove. Label what the parts are made of (plastic? metal? glass?) and whether or not you think they are part of an electronic circuit.

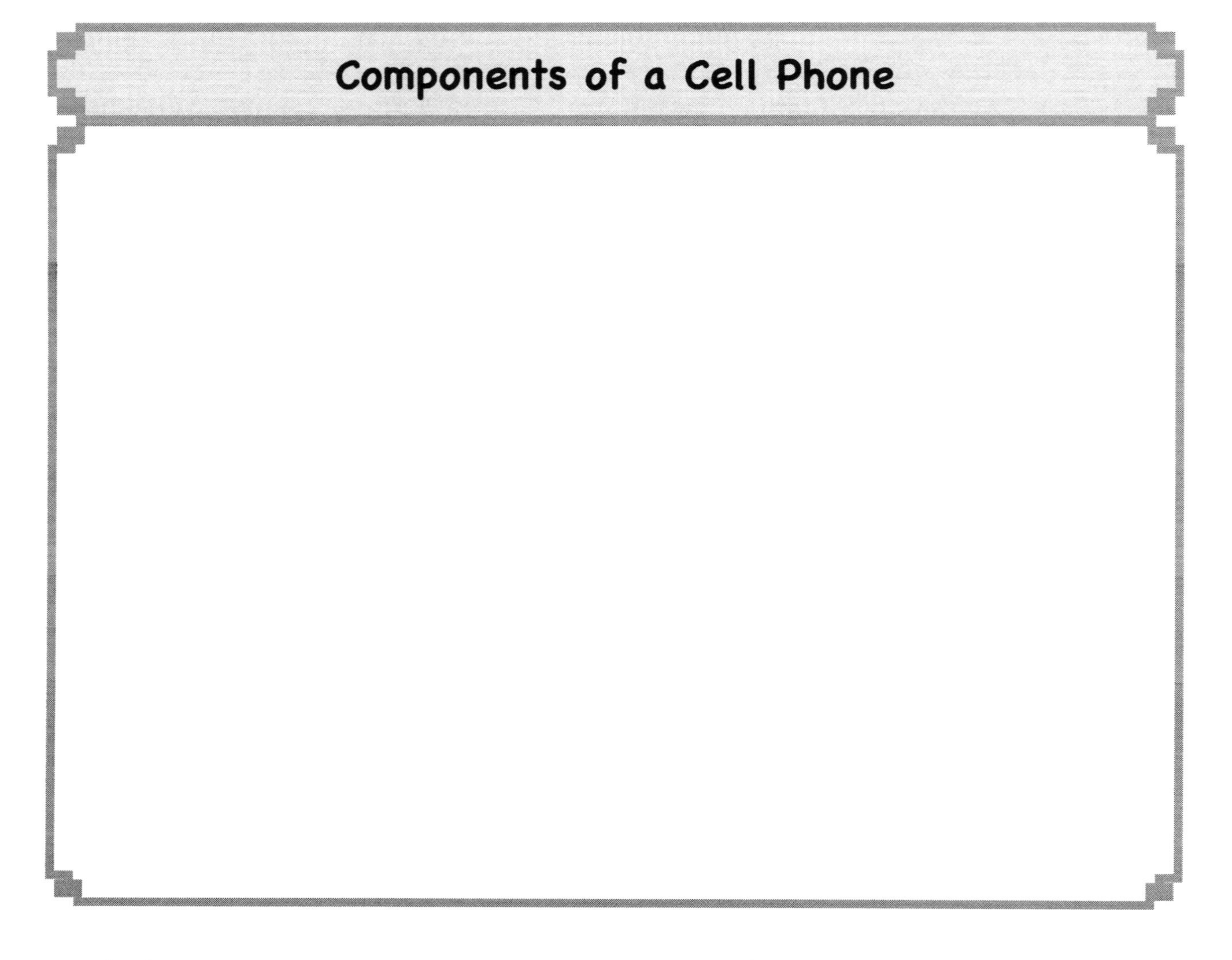

III. What Did You Discover?

❶ How easy or difficult was it to take apart your cell phone?

❷ How old is your cell phone?

❸ List some of the materials used to make your cell phone.

❹ Do you think chemistry was required to create your cell phone? Why or why not?

❺ Do you think physics was required to create your cell phone? Why or why not?

❻ Do you think astronomy is needed for cell phones to work? Why or why not?

---

❼ Do you think biology was required to create your cell phone? Why or why not?

---

❽ Do you think geology is required for your cell phone to work? Why or why not?

---

❾ Do you think the technology behind cell phones would be possible without chemistry, physics, biology, geology, and astronomy together?

IV. Why?

It's easy to forget that all of the technology we use today started with discoveries in basic sciences like chemistry and physics. A cell phone is a modern device that uses discoveries from all areas of science. The materials used to make cell phone casings and electronic components were discovered in the chemistry lab. Physics contributed to our understanding of electrons, electrical circuits, and the energy that powers the cell phone. An understanding of the biology of human hearing, sight, and touch contributed to the design of the cell phone. Knowledge of astronomy made it possible to put satellites in orbit around Earth, and some of these satellites can receive and emit signals directly to and from satellite phones and cell phone towers. Geology has helped cell phone companies find suitable locations for cell phone towers and find the metals used to make cell phones. Without discoveries in chemistry, biology, physics, astronomy, and geology, cell phone technology would not be possible.

V. Just For Fun

Find another electronic device to take apart. Open it and see if you can find any components that are similar to those in the cell phone. Do you find a battery? Electronic circuits? A SIM card?

Carefully remove as many components as you can without destroying them. Wear rubber gloves while taking the device apart.

In the following box, draw the components you removed from the second device. Record your observations about what is similar and what is different between the two devices.

Electronic Device Components

Experiment 2

To Share or Not To Share

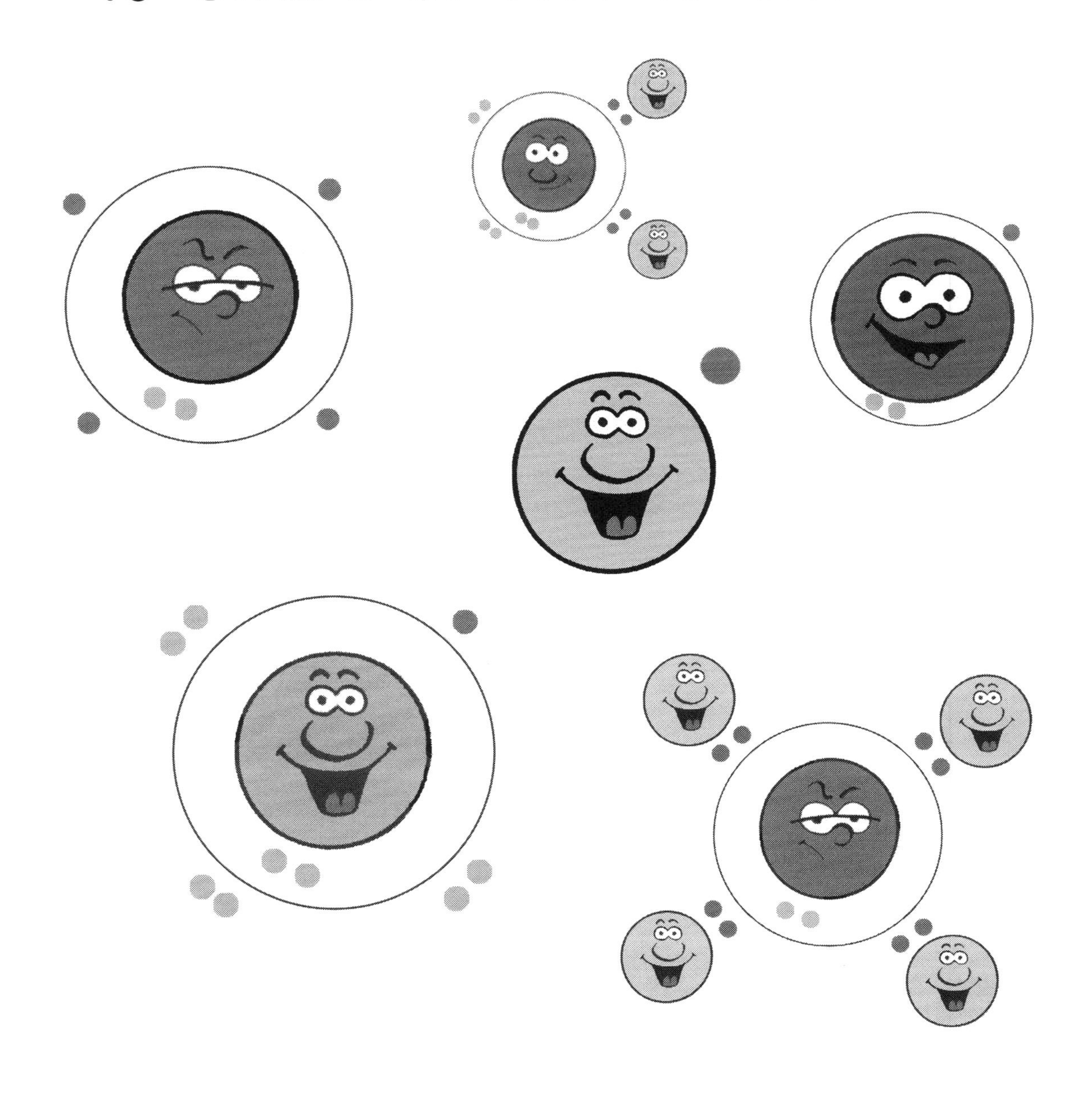

CHEMISTRY

Introduction

Find out how you can model the structure of an atom and the way it transfers electrons during a chemical reaction.

I. Think About It

❶ How are models used in science?

❷ How do models help scientists understand the world around them?

❸ What are some advantages of using models?

4. What are some disadvantages of using models?

____

____

____

____

____

5. How and why do models change?

____

____

____

____

____

II. Observe It

Atoms in molecules like to have their electrons in pairs. Keep this in mind as you make your models.

1. Create a lithium atom.

   Start with a purple jelly bean. The purple jelly bean will represent the center of the lithium atom.

   Lithium has three electrons but there is only one that it likes to give away. To represent these electrons, add 1 red jelly bean and two gray jelly beans to the outside of lithium's center.

❷ Create a fluorine atom.

Start with a green jelly bean for the center of the fluorine atom. Fluorine has 9 electrons, but only one it can use to pair with another electron. Give fluorine 8 gray jelly beans and one red jelly bean.

❸ Create a lithium-fluorine bond to make a lithium fluoride molecule.

Position the lithium and fluorine atoms next to each other. To make a bond between the lithium and the fluorine, position lithium's extra electron so that it is in fluorine's ring of electrons. The rest of the lithium atom will be hanging out nearby.

In the following box, draw your results.

CHEMISTRY

❹ Create 4 hydrogen atoms.

Hydrogen has one electron. It likes to give away its electron or pair it with another electron. Make 4 hydrogen atoms by using a white jelly bean for hydrogen's center and a red jelly bean for the electron.

❺ Create a carbon atom.

Carbon has six electrons, but only four of those are used to create bonds. Use a black jelly bean for the center of the carbon atom. Take 4 red jelly beans, and place one on each of the 4 sides of the carbon. Add two gray jelly beans near the center of the carbon atom.

❻ Create a methane molecule.

Take each of the 4 hydrogen atoms and place them on the four sides of the carbon atom, positioning the red hydrogen electrons so they pair with the red carbon electrons.

Draw your results.

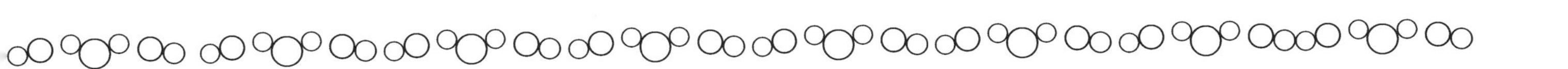

❼ Create a water molecule.

Make an oxygen atom by using a pink jelly bean for the center. You will need 8 electrons, 2 red and 6 gray. Position the electrons around the center by putting three pairs of gray jelly beans around the upper half of the pink jelly bean center. Place one red jelly bean on the lower right side of the center and one red jelly bean on the lower left side.

Make two hydrogen atoms. See Step ❹.

Take the two hydrogens and add them to the oxygen, one hydrogen on each side of the oxygen so they form bonds. This will create a water molecule.

Draw your results.

III. What Did You Discover?

❶ What is the total number of bonding electrons around the fluorine atom in the lithium fluoride molecule in Step ❸?

____

❷ What is the total number of bonding electrons around the carbon atom in the methane molecule in Step ❻?

____

❸ How many bonds are formed in Step ❻, the methane molecule? Are these bonds covalent or ionic? Why?

____

____

____

____

____

❹ What is the total number of bonding electrons around the oxygen in the water molecule?

____

❺ How many bonds are formed in the water molecule?

____

IV. Why?

You will see that most atoms like to have pairs of electrons around them. Fluorine has 5 pairs of electrons around it in the lithium fluoride molecule. Carbon has 5 pairs of electrons around it in the methane molecule. And oxygen has 5 pairs of electrons around it in the water molecule.

However, hydrogen only has one pair of electrons around it. When forming molecules, atoms such as fluorine, oxygen, and carbon are most stable when surrounded by 5 pairs of electrons in a molecule. Hydrogen is most stable when surrounded by one pair of electrons in a molecule.

A molecule that is *stable* is "happy," and although it can still react chemically, it will only do so if the right environment is available. It's like being "satisfied" by a good dinner. You can still eat, but only if the dessert looks really good.

V. Just For Fun

Think of some molecules. Find other colored objects you can use to make models of these molecules, or use more jelly beans.

Draw your molecule models in the space on the following page. Label the atoms and molecules with their names.

Magnificent Molecules

Experiment 3

Three States of Water

Introduction

Pour yourself some water and experiment with getting it to change from one state to another.

CHEMISTRY

I. Think About It

❶ What do you think would happen if you put a cup of water in the refrigerator?

It would stay cold.

❷ What do you think would happen if you put a cup of water in the freezer?

It would turn to ice.

❸ What do you think would happen if you put a cup of water outdoors in the sun (if your dog didn't drink it)?

The water would be warm and eventally evaperate

❹ What do you think would happen if you poured the cup of water into a teapot on the stove and put the heat on high?

It would evaporate into gas.

CHEMISTRY

5. Do you think when you are outdoors you can see water in all its different states? Why or why not?

If the water is first put out it's in liquid form. If it's cold it will turn to ice which is solid form and if it's hot it will turn to gas. So, yes.

II. Observe It

1. Pour some water into an ice cube tray or paper cup. Use your senses to test the water. What can you observe about the water?

It's a liquid and clear.

2. Put the tray or cup of water in the freezer. Check periodically to see if the water has frozen solid—or leave it overnight. Once the water has frozen, use your senses to test it. What can you observe about the water?

It's frozen and cold to the touch. There's little crystals growing on it.

3. Take the ice out of its container and put it in a small frying pan or saucepan. Put the pan on a stove and have your teacher help you turn the stove to low or medium heat. Without touching it, what can you observe about the ice as the stove warms up?

It's turning into water and melting.

❹ Have your teacher help you bring the water to a boil and then place a lid on the pan. Leave it for a few minutes. Next, take the lid off the pan and look at the inside of the lid. What can you observe?

It turned into water

III. What Did You Discover?

❶ Did you observe the water in all three states? Why or why not?

I did. I saw steam which is gas.
I saw ice which is a solid
I saw water which is a liquid.

❷ Did you observe the water changing from one state to another? Why or why not?

Yes. I saw water turn to ice, ice turn to water, and water turn to gas.

❸ When you took the lid off the pan and looked at the inside of the lid, did you see water there? Why do you think this happened?

I saw water. I think the heat melted the ice.

❹ When you heated the water, was there as much water in the pan at the end as when you began? Why do you think this happened?

No. I think the heat made the water go crazy so it floated upwards.

CHEMISTRY

IV. Why?

Water is amazing! We use water in all three of its states every day. In its liquid form we drink it, cook with it, bathe in it, swim in it, and use it to grow plants. In its gaseous form we breathe it in as water vapor in the air, and we might go to the sauna for a steam bath. In its solid form, we use it to cool our drinks, it falls from the sky as snow, and we can skate on it in the winter.

When you heated the ice, the heat energy from the stove made the water molecules in the ice move faster. This made the lattice structure break apart, and the water turned back to its liquid state. As you continued to heat the water and it began to boil, you probably noticed steam coming from the water. The steam is water evaporating and entering the air in its gaseous state. This is why you may have noticed that you had less water at the end of the experiment than you did when you began.

You probably noticed water on the inside of the lid when you took it off the pan of boiling water. This happened because the water *condensed* on the lid. The water vapor evaporating from the boiling water was very hot with fast moving molecules. When the hot water vapor hit the cooler pot lid, the molecules gave some of their heat to the lid which caused the molecules to move more slowly. The water vapor then turned back to liquid water.

A change in the state of water is not caused by a chemical change since the water molecules are still the same in each state. When water undergoes a change from one state to another, it is called a *physical change*. Just the physical properties of the water have changed, but the molecules stay the same chemically.

V. Just For Fun

Do you think cold water or hot water will freeze faster?

Direct your teacher to help you with this experiment.

❶ Have your teacher boil some water.

❷ Take a Styrofoam cup. Fill it about half full with cold tap water.

❸ Have your teacher put boiling water of the same amount in a second Styrofoam cup.

❹ Put your cup of cold water in the freezer and have your teacher put their cup of boiling water in the freezer next to it.

❺ Check the two cups of water frequently to see which freezes faster.

❻ Record your observations.

Experiment 4

Fast or Slow?

Introduction

Observe whether having more or less heat makes a difference in a chemical reaction.

I. Think About It

1. Do you think adding heat could make a chemical reaction happen faster? Why or why not?

   I think so. Heat makes things go faster so I don't think its different with reactions.

2. Do you think taking away heat could slow down a chemical reaction? Why or why not?

   Yes. The less heat the slower things happen.

3. Do you think atoms and molecules can be made to move faster? Why or why not?

   Well using heat makes them go crazy so yes.

4. Do you think it makes a difference for a chemical reaction if atoms and molecules are moving fast or slow? Why or why not?

   Yes, if the molecules are slow the chemical reaction will happen slower, and if fast, faster.

CHEMISTRY

❺ Do you think the speed of a chemical reaction will be different if there are more molecules or less molecules? Why or why not?

II. Observe It

❶ Take 2 Styrofoam cups and measure 60 milliliters (1/4 cup) of apple cider vinegar into each cup.

❷ Measure 120 milliliters (1/2 cup) of cold tap water and pour into a different Styrofoam cup.

❸ Add 15 milliliters (1 tablespoon) of baking soda to the cold water and stir to dissolve it.

❹ Have your teacher measure 120 milliliters (1/2 cup) of boiling water and pour it into a fourth Styrofoam cup.

❺ Carefully add 15 milliliters (1 tablespoon) of baking soda to the boiling water and stir to dissolve it.

❻ Pick up both cups of vinegar. At the same time, pour each into a cup of baking soda water.

❼ On the next page, record your observations.

CHEMISTRY

Observations of Reaction with Hot Water

Observations of Reaction with Cold Water

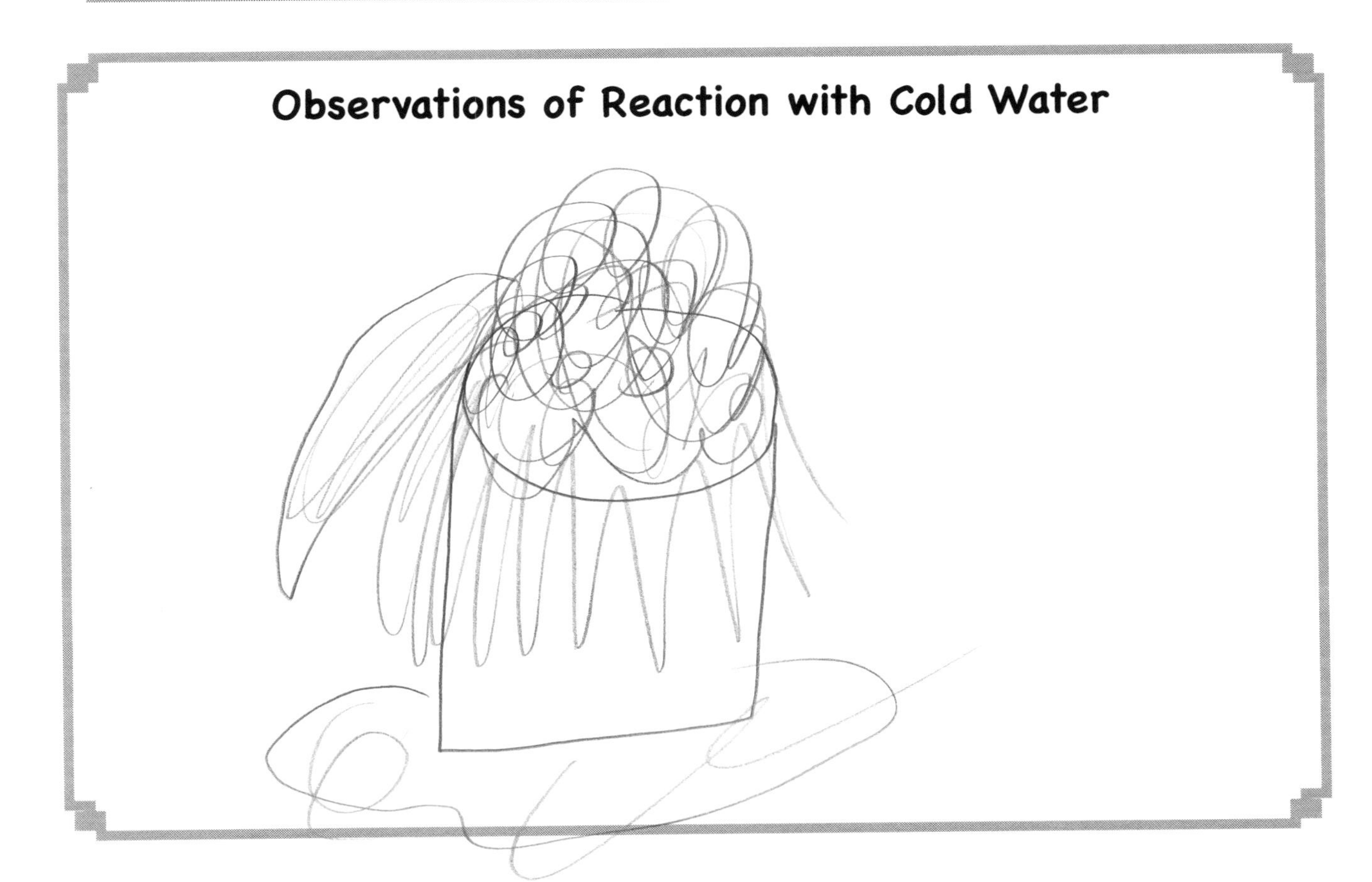

CHEMISTRY

III. What Did You Discover?

1. Was it harder or easier to dissolve the baking soda in the hot water than the cold water? Why?

It was easier because the heat made it go faster

2. Did the reaction happen faster with one baking soda mixture than the other? If so, which one? Why?

The reaction happened faster with hot water. Apparently, hot makes it go faster.

3. Did anything happen that was different between the hot and cold baking soda water reactions? Why or why not?

The cold water went crazy while the hot water was not as crazy.

4. Did anything happen that was the same between the hot and cold baking soda water reactions? Why or why not?

They both fizzed.

IV. Why?

Recall that heat speeds up chemical reactions. When the vinegar is added to the boiling baking soda water, the reaction goes more quickly than with the cold baking soda water. The molecules in the boiling water are moving much more rapidly than the molecules in the cold water. The faster moving molecules in the boiling water have more energy to give to the reaction, causing it to happen faster. The slower moving molecules in the cold water have less energy to give to the reaction, and so it happens more slowly.

V. Just For Fun

Repeat the experiment. This time use balsamic vinegar instead of apple cider vinegar. Do you notice any difference from the first experiment? Record your results.

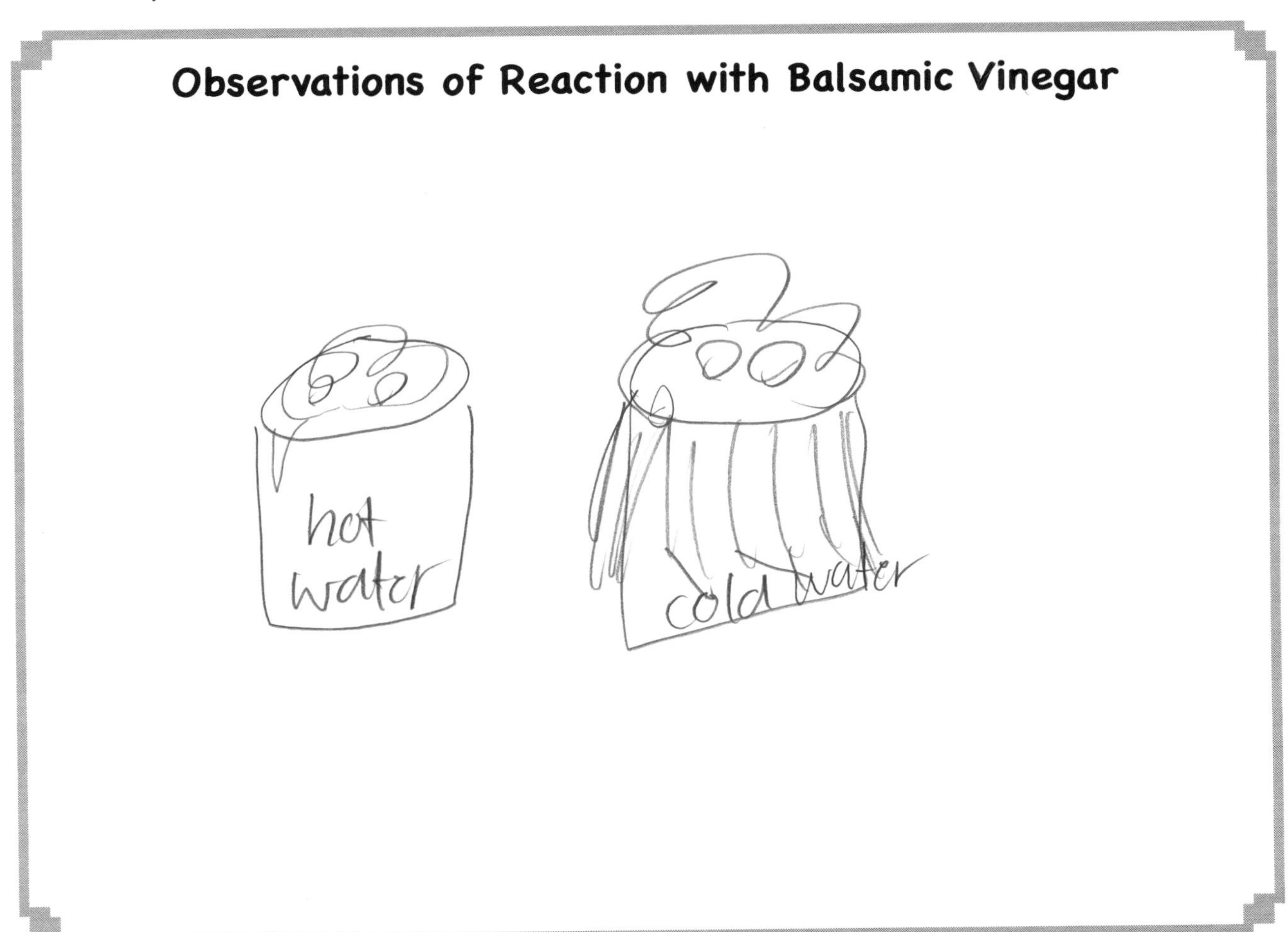

Experiment 5

Eggs—Hot or Cold?

CHEMISTRY

Introduction

Find out whether adding heat and taking away heat from scrambled eggs causes a change of state or a chemical reaction.

I. Think About It

1. If you beat eggs with a fork, do they look different than before you beat them? Why or why not?

____

____

____

____

2. When you beat eggs with a fork, do you think this causes a chemical reaction? Why or why not?

____

____

____

____

3. Do you think cooking eggs (adding heat) causes a chemical reaction? Why or why not?

____

____

____

____

❹ Do you think freezing uncooked eggs causes a chemical reaction? Why or why not?

____

____

____

____

❺ Do you think freezing cooked eggs causes a chemical reaction? Why or why not?

____

____

____

____

II. Observe It

❶ Take two raw eggs. Break each into a separate bowl and scramble each with a fork.

❷ Put one of the scrambled eggs in a plastic bag, seal the bag, and put it in the freezer.

❸ Take the other scrambled egg and have your teacher help you cook it on the stove. Set the cooked egg aside to cool.

❹ When the egg in the freezer is frozen, take it out of the freezer and let it thaw without heating it.

5. Once the egg has thawed, compare it to the cooked egg.

6. Record your observations.

Cooked Egg/Thawed Egg Comparison

7. Cook the egg that was frozen and compare it to the first cooked egg.

8. Record your observations.

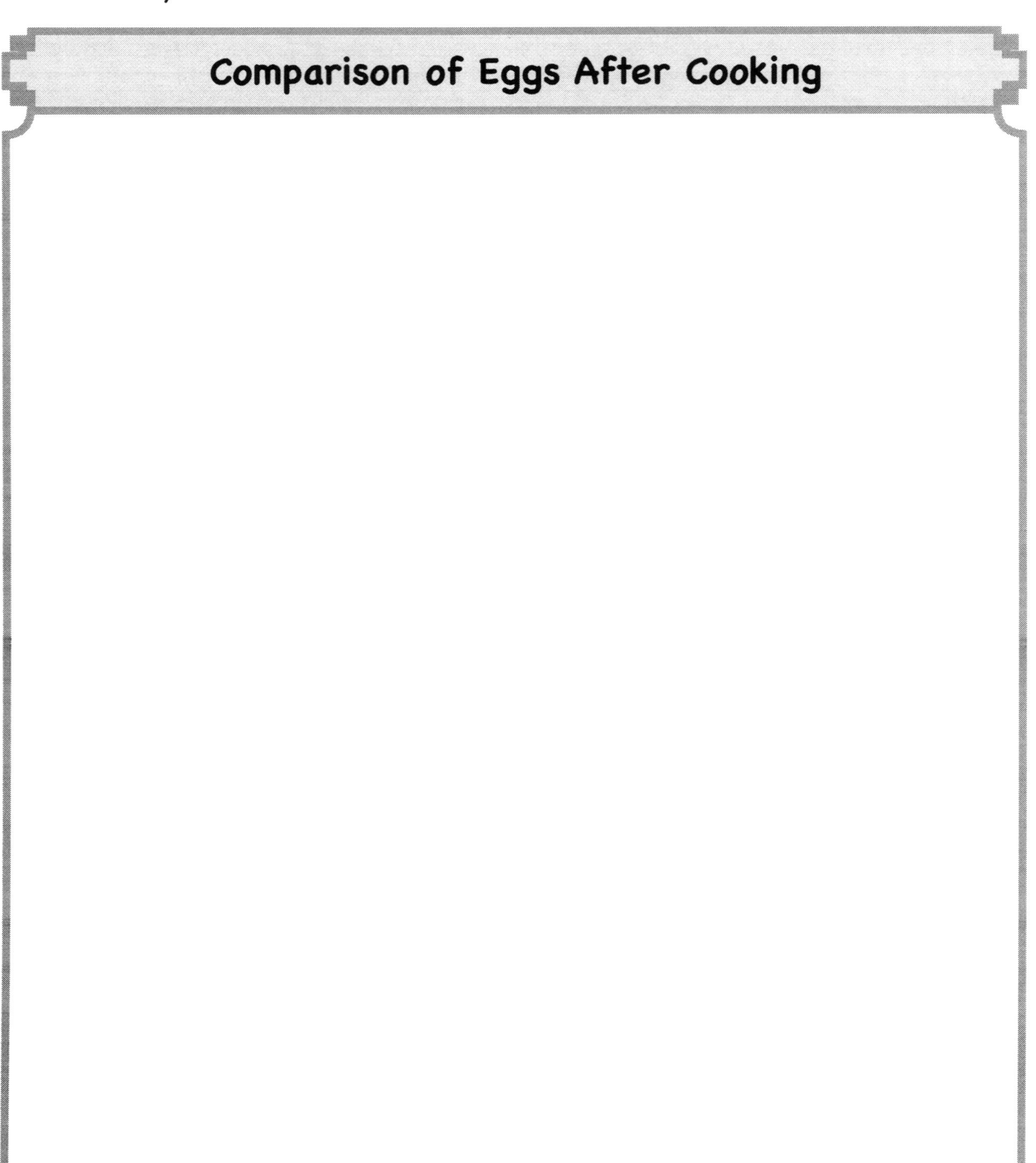

CHEMISTRY

III. What Did You Discover?

❶ What differences did you observe between heating and freezing eggs?

❷ When you cooked the egg in Step ❸, did it undergo a chemical reaction? Why or why not?

❸ When you thawed the frozen egg, could you tell if freezing it had caused a chemical reaction? Why or why not?

❹ What happened when you cooked the egg that had been frozen? Was it different from cooking the egg that hadn't been frozen? Why?

IV. Why?

Adding or taking away heat can cause chemical reactions to happen and can also cause changes of state. However, chemical reactions and changes of state are different ways that heat can be transferred. Adding heat to a raw egg by cooking it causes a chemical reaction in which bonds break and form new bonds. When the egg is cooked, you can't un-cook it. That is, it doesn't go back to being a raw egg after you stop adding heat.

However, if you freeze a raw egg, a change of state occurs but no chemical reaction occurs. A raw egg will go from a liquid state to a solid state when it is frozen, but because no bonds are broken and reformed, when the raw egg is thawed, it goes back to a liquid state.

V. Just For Fun

Take an ice cube and place it in a plastic bag. Seal the bag.

Hold the ice cube in the plastic bag in your hands. Observe what happens. How is heat transferred?

Record your observations.

Ice in Hands

Experiment 6

Nature Walk

Introduction

Why is observing animals the most important way to learn about them? Go on a nature walk to find out!

I. Think About It

❶ What types of animals have you observed?

❷ Do you have animals that live in your house? What are they and how would you describe them?

❸ Do you have animals that live in your yard? What are they and how would you describe them?

❹ Have you ever seen animals at a zoo? If so, what kinds? How would you describe them? How are they the same and how are they different?

____

____

____

____

____

____

❺ What features of animals have you observed that are common to many animals?

____

____

____

____

____

____

❻ What unique features let you know the difference between the kinds of animals you are observing?

____

____

____

____

____

____

II. Observe It

Scientists keep notebooks where they can write down observations they make while they're studying living things outdoors. These are called *field notebooks*. In this experiment you'll make your own field notebook.

1. Take your notebook, a pencil, and some colored pencils and put them in a backpack along with a snack and some water.
2. Find a wooded area, park, or other place near your home where you can observe animals. Or you might take a trip to the zoo.
3. Locate several animals you would like to study. Notice what features each animal has. Write the names of these animals and their features in your field notebook.
4. Observe what each animal is doing. Are they moving, eating, sleeping, communicating, or watching you? Where are they—on the ground, in a tree, in a pond, or someplace else? What else can you observe? Write down your observations.
5. Now make a quick drawing of each animal. You can record their shape and color, what their ears and tail look like, and any other features you notice.
6. If you have a camera, take a photo of each animal, print it, and place it in your notebook alongside your drawing.
7. Plan some other times when you can go for a walk and add more animal observations to your field notebook. Now you are a biologist!

III. What Did You Discover?

❶ Which animals did you observe? Where were they?

❷ Did the animals move? If so, how and how much?

❸ Did you observe any animals that did not move? If so, why were they not moving? Where were they?

BIOLOGY

❹ Were any of the animals eating? If so, what were they eating and how were they eating it?

____

____

____

____

____

____

❺ Were any of the animals communicating? If so, how?

____

____

____

____

____

____

❻ What did you observe about animals that you hadn't noticed before?

____

____

____

____

____

____

IV. Why?

Observing animals in their environment is the first step to learning about them. By continuing to make careful observations, you will know what the different animals look like, what color they are, what their skin or hair or outer shell looks like, how large or small they are, and what kinds of food they eat. You can learn how different kinds of animals interact with each other and depend on each other.

Keeping a field notebook allows you to keep all your observations in one place so you can refer to them later. You can add more information about each animal as you make new observations, and you can add more kinds of animals to observe. When you observe animals over a long period of time, you learn about how they grow, how their activities change with the seasons, and how they interact, along with many other interesting details.

V. Just For Fun

Keep adding to your field notebook. Observe the animals in your area for several months. How do they change—or do they stay the same? Find some additional animals to observe. Write about, draw, and photograph them. (Did you remember that bugs are animals?)

Experiment 7

Red Light, Green Light

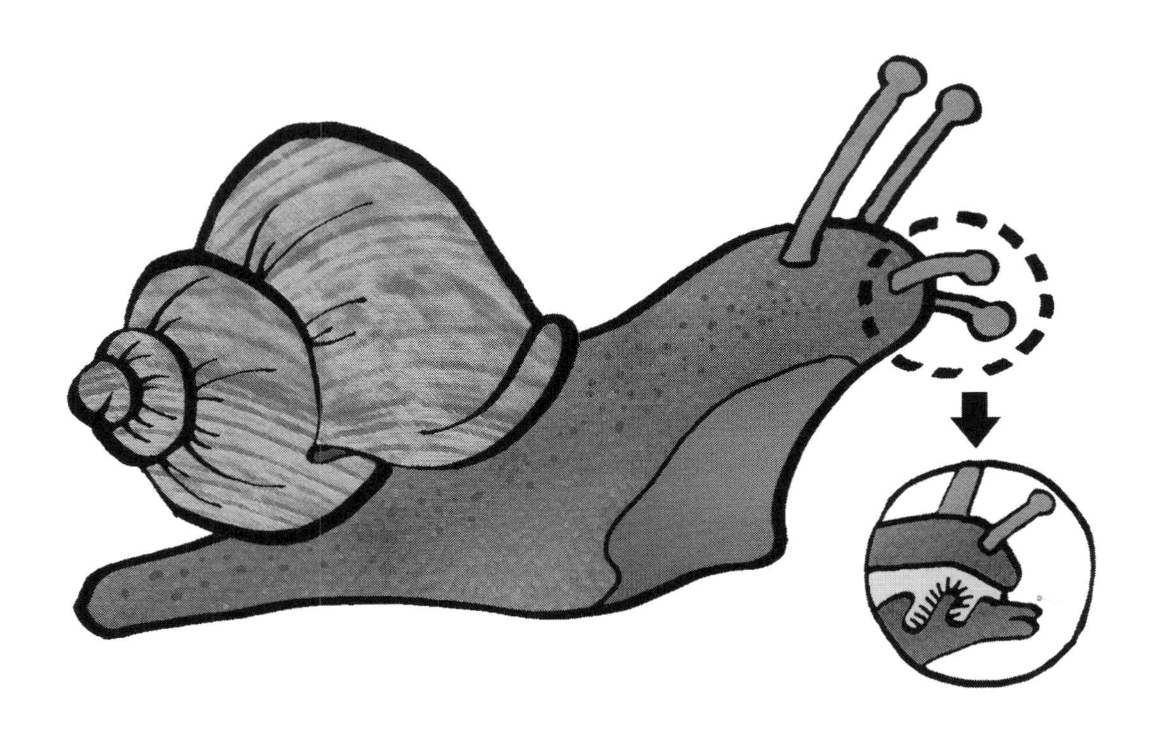

Introduction

Do you think snails and earthworms will move across any surface? Test it!

I. Think About It

❶ What do you think snails like?

____

____

____

____

____

❷ What do you think snails don't like?

____

____

____

____

____

❸ What do you think earthworms like?

____

____

____

____

____

4. What do you think earthworms don't like?

5. If you want to keep snails out of your garden, how would you do it?

6. If you want to encourage earthworms to stay in your garden, how would you do it?

II. Observe It

❶ Take a large plastic box or tray and add enough garden dirt to cover the bottom.

❷ Take some water and moisten the garden dirt in one-half of the box (one end). Use just enough water to make the dirt moist. The other half of the dirt needs to stay somewhat dry.

❸ Perform a control experiment by placing the earthworms and/or snails on the dry side of the box. Observe whether or not they move to the moist end. Record your observations below.

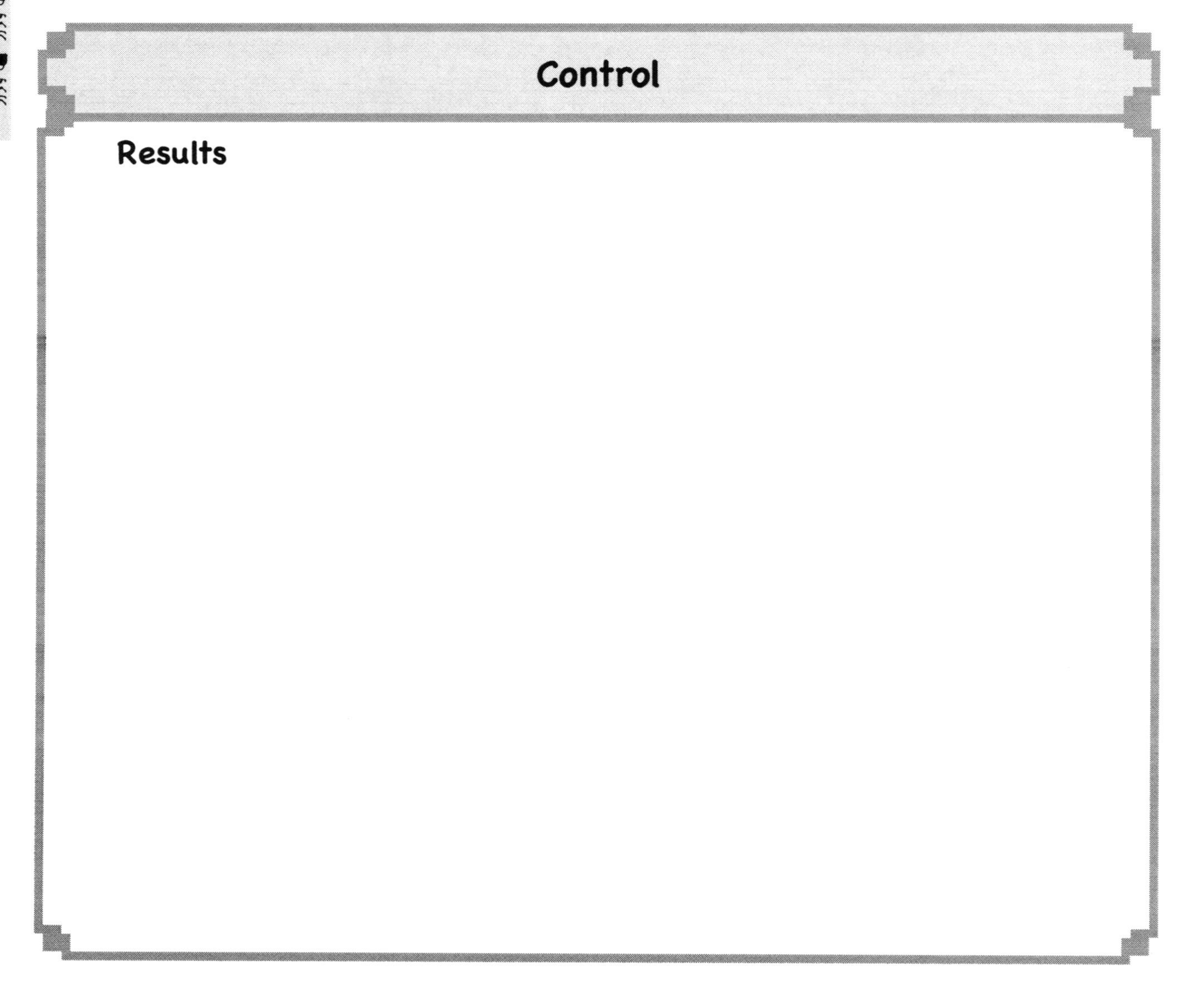

4. Carefully remove the snails and/or earthworms and place them back in their holding box.

5. Choose one of the powders and pour it in a line across the soil from one side of the box to the other about halfway between the two ends of the box.

6. Place the earthworms and/or snails on the soil in the dry end of the box and observe whether or not they cross the powder barrier.

7. Record your observations.

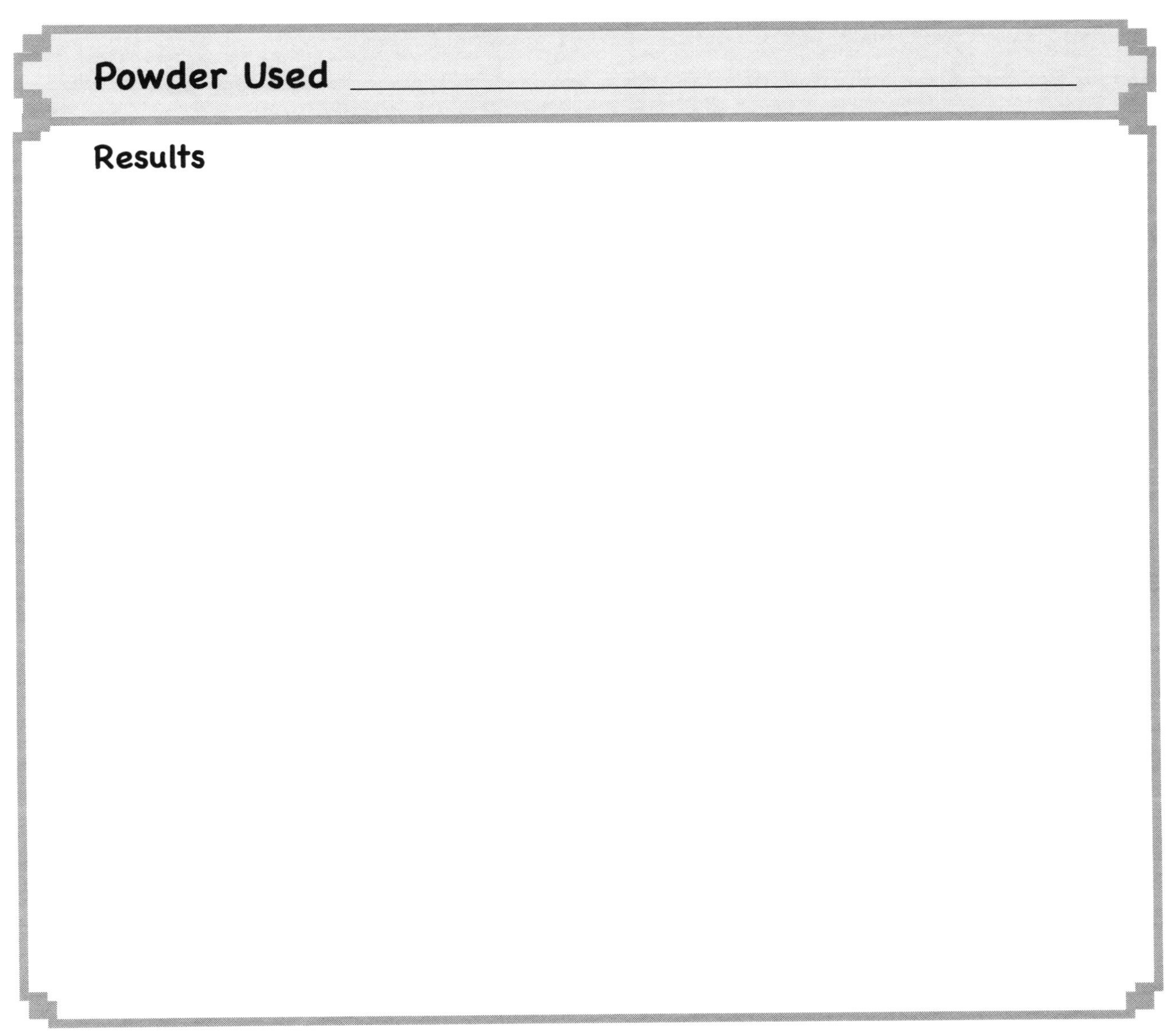

❽ Gently remove the snails and/or earthworms and place them back in their holding box.

❾ Remove the powder and fill in this area with new dirt, then pour a different powder in a line across the dirt. Repeat Steps ❻–❼

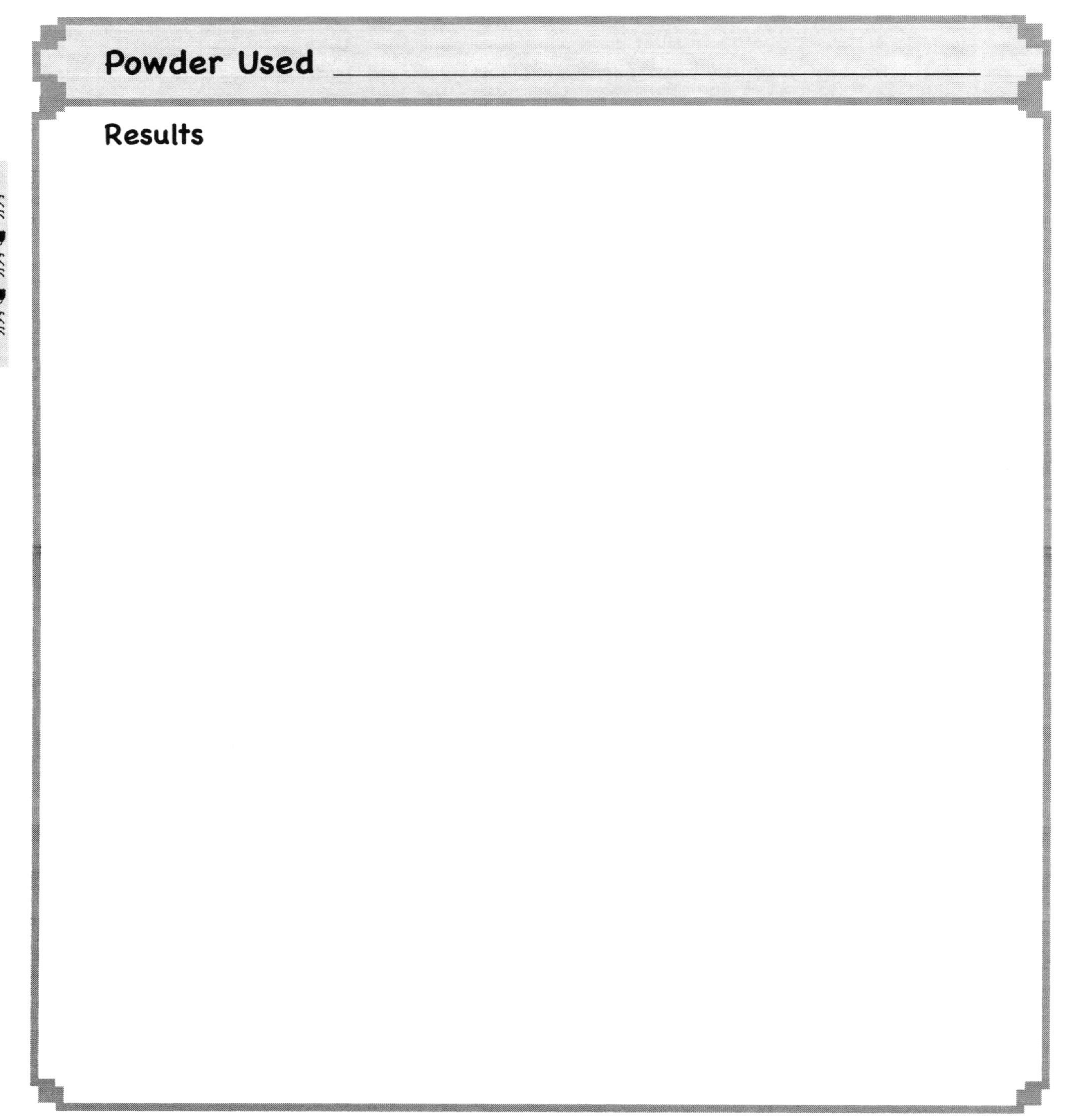

⑩ Repeat Steps ❽–❾ using two different powders.

Powder Used ______________________________

Results

Powder Used ______________________________

Results

III. What Did You Discover?

1. When there was no barrier, did the snails and/or earthworms move from the dry end to the moist end of the box? Why or why not?

2. Were there any powders that the snails and/or earthworms would not cross? Why or why not? If there were any barriers they wouldn't cross, list them.

3. Use the table below to chart which of the powder barriers the snails and/or earthworms crossed and which they would not cross.

Powder	Snails WOULD cross	Snails would NOT cross	Worms WOULD cross	Worms would NOT cross

❹ Could you use this experiment to find ways to keep snails out of your garden but earthworms in your garden? How would you do it?

____

____

____

____

____

IV. Why?

Worms and snails are not as smart as octopuses, but they can choose good environments, and they have behaviors that help them escape from birds so they won't be eaten. They have the ability to know if some substances are dangerous to them and then to avoid those substances.

Snails can withdraw into their shell for protection, and they hide in dark moist places, like under rocks and foliage. Because snails need to stay moist in order to live, they usually hide during the day. Their bodies can dry out and die in the hot sun. Hiding during the day also protects them from predators. If the weather is cloudy or rainy, you may see snails out during the day, and they are often out in the cooler times of dawn and dusk and at night. When snails come out, they eat a lot! They can be very harmful to a garden, destroying many plants.

Earthworms are great to have in your garden because they help plants grow. Like snails, earthworms need to be in a moist environment or they will dry out and die. To escape predators, earthworms spend most of their time underground. You might see earthworms on the sidewalk after a big rain, but scientists are still not sure why earthworms come out when it rains. As a result of doing experiments, scientists think that earthworms communicate with each other through touch and that earthworms sometimes travel in groups.

V. Just For Fun

Do you think you can do a similar experiment to find out how to keep ants out of your house? Find an active anthill to use in your experiment.

Write the steps you'll take to perform your experiment:

Perform your experiment and record your observations on the next page.

Results of Ant Experiment

Experiment 8

Butterflies Flutter By

Introduction

This experiment is about observing the life cycle of an arthropod. A *cycle* is a series of events that repeat, and a *life cycle* is the series of events from birth through death of an animal.

I. Think About It

❶ How did the butterfly get its name? Is it because butterflies eat butter? Or do you think it is that butterflies are sometimes yellow—like butter?

Write or draw how you think the butterfly got its name.

❷ How does a caterpillar turn into a butterfly? Draw or write how you think a caterpillar becomes a butterfly.

II. Observe It

❶ **The Beginning: The Egg**

Draw and/or write what you see.

❷ The Middle: The Caterpillar

Draw and/or write what you see.

❸ The Change: The Chrysalis

Draw and/or write what you see.

❹ The End: The Butterfly

Draw and/or write what you see.

❺ Draw the life cycle of the butterfly as you observed it.

BIOLOGY

❻ What else did you notice that was very interesting? Draw it below.

III. What Did You Discover?

1. Did the butterfly eggs look like you thought they would? Why or why not?

2. Did the caterpillar look like you thought it would? Why or why not?

3. Were you able to observe exactly what was happening in the chrysalis? Why or why not?

4. Did the butterfly look the way you thought it would? Why or why not?

❺ Describe your favorite part of the butterfly life cycle. Explain why it is your favorite part.

____

____

____

____

IV. Why?

The butterfly is a creature that starts as something completely different—a caterpillar. If you had never watched a butterfly egg turn into a caterpillar and a caterpillar turn into a butterfly, you would not know that they are the same creature. It takes a keen eye and careful observation to find out what the life cycle is for the butterfly.

How did the butterfly get its name? You probably found out that no one is quite sure. There are ideas about how the butterfly got its name, but not everyone agrees. One idea is that the word butterfly comes from a very old word *buturfliog* which is a word coming from "butter" and "fly." But why butter? One idea is that butterflies like to eat butter and land on creamy, buttery foods in kitchens. The German word for butterfly means "milk thief," so maybe butterflies like butter and milk. But no one is really sure where the name for butterfly came from.

Sometimes disagreements happen even in science. We know much about the life of a butterfly because we can observe it. But scientists don't know everything because scientists can't observe everything. You were not able to observe exactly what was going on inside the chrysalis as the butterfly changed, because you couldn't see inside the chrysalis—but you were able to make some observations. The most important job of scientists is to make careful observations and then to record exactly what they see, even if they can't see everything—just like you did in this experiment!

V. Just For Fun

Using your *Student Textbook* as a reference, write some features of arthropods in your field notebook. Take your field notebook and go outside to look for arthropods. Use the features you have written down to help determine if the creature you're observing is an arthropod.

Also look for arthropods in different stages of their life cycle. Can you find eggs? Larvae? (Larvae are the newly hatched form of some insects. For example, caterpillars are larvae.) Chrysalises? Young arthropods? Adult arthropods?

In your field notebook, make notes and drawings about what you observe.

Experiment 9

Tadpoles to Frogs

Introduction

Discover the different stages in a frog's life cycle.

I. Think About It

❶ How did the frog get its name? Is it because frogs sit on logs? Or do you think it is that frogs have deep voices? Write or draw how you think the frog got its name.

❷ What do you think will happen as a tadpole turns into a frog? Write or draw what you think.

II. Observe It

❶ The Beginning: The Egg
Draw and/or write what you see.

❷ The Middle: The Tadpole Eating

Draw and/or write what you see.

❸ The Change: The Tadpole with Hind Legs

Draw and/or write what you see.

❹ The Change: The Tadpole with Front Legs

Draw and/or write what you see.

❺ The End: The Adult Frog

Draw and/or write what you see.

BIOLOGY

❻ Draw the life cycle of the frog as you observed it.

❼ What did you notice that you thought was very interesting? Draw it below.

III. What Did You Discover?

❶ Did the frog eggs look like you thought they would? Why or why not?

____

____

____

____

____

____

❷ Did the tadpole look like you thought it would? Why or why not?

____

____

____

____

____

____

❸ Were you able to observe exactly what was happening when the tadpole started to change ? Why or why not?

____

____

____

____

____

____

4. Did the adult frog look like you thought it would? Why or why not?

____

____

____

____

____

____

5. Describe your favorite part of the frog life cycle. Explain why it is your favorite part.

____

____

____

____

____

____

IV. Why?

A frog starts life as an egg and then becomes a tadpole before it becomes an adult frog. It changes significantly during its life cycle. A tadpole doesn't look like a frog but more like a fish. Even though a tadpole may look like a fish, it is not a fish. If you never observed a tadpole changing into a frog, you wouldn't know they are the same creature.

Some creatures like frogs and butterflies undergo a drastic change in appearance when they become adults. This process is called *metamorphosis*. Metamorphosis simply means "to change form or shape."

Humans do not undergo a metamorphosis as they grow into adults. Although you will look different when you are an adult than you do now, you do not completely change your form or your shape as you grow. You may get taller and your body will change proportions, but overall you keep the same shape and form you had when you were born. You are a different kind of creature than a frog or a fish!

V. Just For Fun

Go outside and look for frogs and toads, lizards and snakes, and all kinds of birds. What can you observe about where they are in their life cycle? Do you see any eggs? Babies? Adults? How many different kinds of frogs, toads, lizards, and snakes can you see? If you're near water, look for fish too!

Take your field notebook with you and make notes and drawings about what you observe.

Experiment 10

Making Waves

Introduction

Use a rope or string to find out more about waves.

I. Think About It

1. What do you think would happen if you dropped a pebble in a pond? Do you think waves would form? Why or why not?

2. What do you think would happen if you dropped a boulder in a pond? Do you think the waves would be bigger than the waves created by a pebble? Why or why not?

3. What do you think would happen if you wiggled a rope on the ground? Do you think you would see the rope form waves? Why or why not?

4. What do you think would happen if you wiggled the rope faster? Would the waves be smaller or bigger, shorter or longer?

____

____

____

____

II. Observe It

1. Crush the colored chalk until it is a powder.
2. Spread the chalk powder on the sidewalk or other flat surface.
3. Carefully place the rope or string so it goes in a straight line through the middle of the chalk dust without disturbing the chalk dust.
4. Lift one end of the rope. Is the chalk dust the same or has it been disturbed?
5. In the space below, record what you observe.

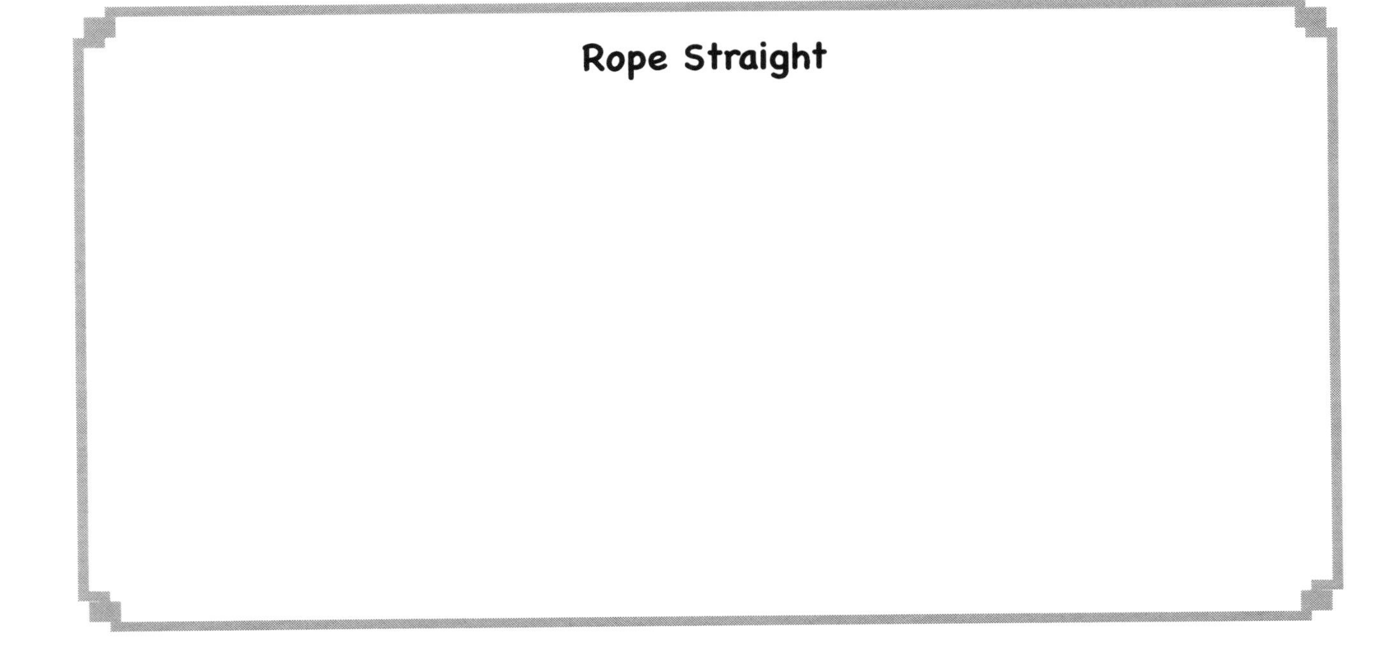

6. Smooth out the chalk dust. Place the rope across the middle of the chalk dust and gently wiggle one end, creating a wave. Lift the rope. Is the chalk dust the same or has it been disturbed?

7. Record what you observe.

❽ Smooth out the chalk dust. Place the rope through the middle of the chalk dust and vigorously wiggle one end, moving the rope as fast as you can. Lift the rope. Is the chalk dust the same or has it been disturbed?

❾ Record what you observe.

III. What Did You Discover?

❶ What happened the first time you placed the string in the middle of the chalk dust? Did you observe a wave or a straight line where the string was located?

❷ What happened when you placed the string in the middle of the chalk dust and wiggled one end? Did you observe a wave or a straight line where the string was located?

❸ What happened when you placed the string in the middle of the chalk dust and wiggled one end vigorously? Were the waves smaller, larger, shorter, or longer than the previous time?

❹ Do you think you created a transverse, longitudinal, circular or electromagnetic wave? Why?

____

____

____

____

IV. Why?

When you place a string in the middle of a patch of colored chalk dust, you can observe what happens when you create a wave. As you hold one end of the rope and wiggle it, the energy you transfer to the rope travels down the rope to the other end. Energy is also transferred to the surrounding chalk dust, moving the dust to create a pattern.

If you wiggle the rope gently, you can observe the waves created by the rope as the chalk dust is scattered. When you wiggle the rope vigorously, you generate shorter wavelengths, but larger amplitudes. If you wiggle the rope more slowly, you generate longer wavelengths, and shorter amplitudes.

V. Just For Fun

Crush several different colors of chalk. Use the chalk dust to make stripes of color on a flat surface. Repeat the experiment and see what kinds of colorful patterns you can create with waves.

Experiment 11

Musical Glasses

Introduction

Discover how sound travels through water and air.

I. Think About It

1. What do you think will happen if you tap an empty glass with a wooden stick?

___

___

2. What do you think will happen if you tap a glass full of water with a wooden stick?

___

___

3. Do you think the sound will be different if you tap an empty glass with a wooden stick and then tap a glass full of water with the wooden stick? Why or why not?

___

___

___

___

4. Do you think you could make different sounds with several glasses, each containing a different amount of water? Why or why not

___

___

___

___

II. Observe It

❶ Take 5 glasses or glass jars. Using a wooden stick, tap each glass lightly. Record what you observe.

Observations of the Sound of Empty Glasses

❷ Fill one glass all the way to the top with water. Fill another glass 2/3 full, another half full, another 1/3 full, and leave the last one empty.

❸ Take a wooden stick and tap each of the glasses. Record what you observe.

Observations of the Sound of Differently Filled Glasses

❹ Tap the glasses several times until you are familiar with the sound each one makes.

❺ Close your eyes or wear a blindfold. Have a friend rearrange the order of the glasses. Then have your friend tap a glass. Guess which glass they are tapping. Have your friend record which glass they tapped and what your guess was. Repeat several times.

Glass Tapped	Glass Guessed

III. What Did You Discover?

1. When you tapped the empty glasses, did they sound different from each other? Why or why not?

2. When you tapped the glasses that had different amounts of water in them, did they sound different from each other? Why or why not?

3. Did the amount of water in a glass affect the sound it made when you tapped it? Why or why not?

4. Which glass had the highest pitched sound?

❺ Which glass had the lowest pitched sound?

____

____

❻ When your eyes were closed, were you able to tell which glass was tapped? Why or why not?

____

____

____

____

IV. Why?

If you used identical empty glasses or jars, when you tapped each one, the sounds were the same. However, when you tapped glasses that contained different amounts of water, the pitch of the sound changed.

When you tap a glass, the sound you hear is the vibration of the glass molecules. The molecules in the glass move, or *vibrate*, with a certain frequency. If the glass is empty, the vibrating glass molecules transfer their energy directly to the surrounding air molecules, and these air molecules carry the sound wave to you ear. If the glass is filled with water, the vibrating glass molecules transfer their energy to the water molecules in the glass *before* that energy is transferred to the air.

Recall that a high-pitched sound has a high frequency and a low-pitched sound has a low frequency. Because a glass full of water has more molecules in it (contains more mass) than a glass full of air, it takes more energy to make the molecules move. Also, some of that energy is lost as the water molecules bump around. This loss of energy is called *dampening*. When energy is lost, the frequency of the sound wave decreases and this lowers the pitch.

V. Just For Fun

Play a game of Marco Polo with a group of friends.

Find a flat, open space. Have your friends spread out. Close your eyes or put on a blindfold and then yell "Marco!" Your friends will answer you by yelling "Polo!"

Catch one of your friends by listening to the sound of their voice. You can continue to yell "Marco!" with your friends answering "Polo!" until you touch one of them.

Observe how easy or difficult it was to find one of your friends by using the sound of their voice to locate them. Do you think sound can be used as a location device? Record your observations and ideas in the following box..

PHYSICS

Experiment 12

Separating Light

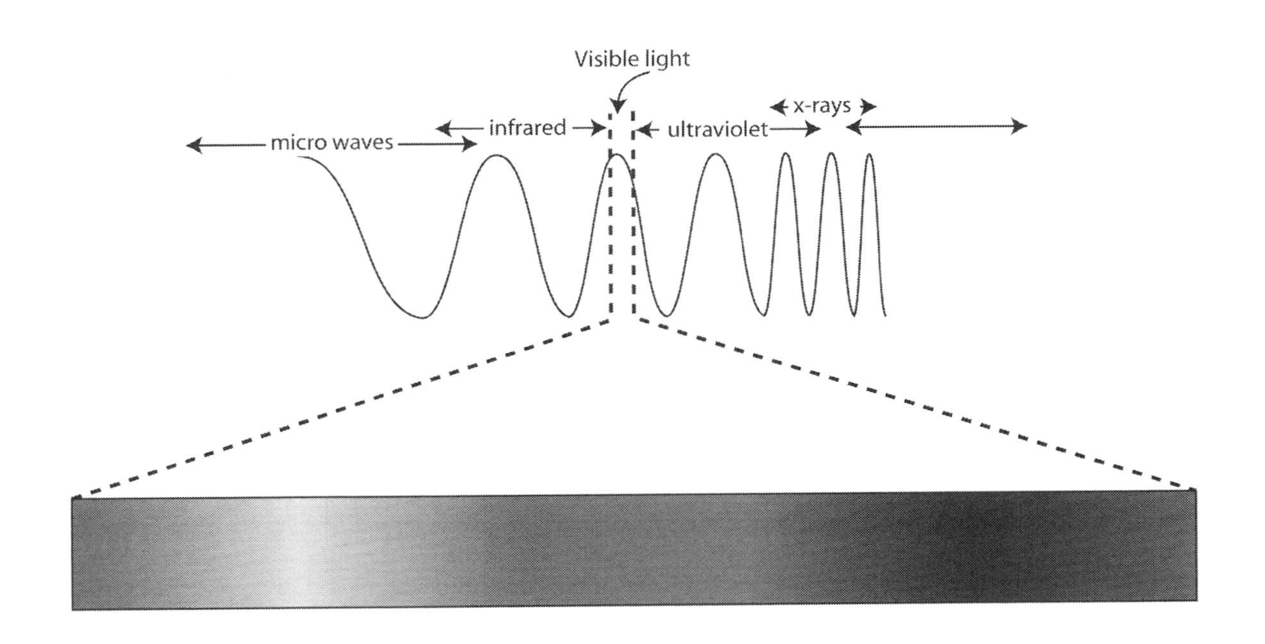

Introduction

Discover how bending white light can separate it into different colors.

I. Think About It

❶ What are the different colors of the rainbow?

❷ Do you ever see yellow, purple, red, and green in that order in a rainbow? Why or why not?

❸ When do rainbows occur? Why?

❹ Do rainbows always occur when it rains? Why or why not?

II. Observe It

❶ Take one prism and shine the flashlight beam through it at the 90° bend. Have a white wall, whiteboard, or white paper behind the prism.

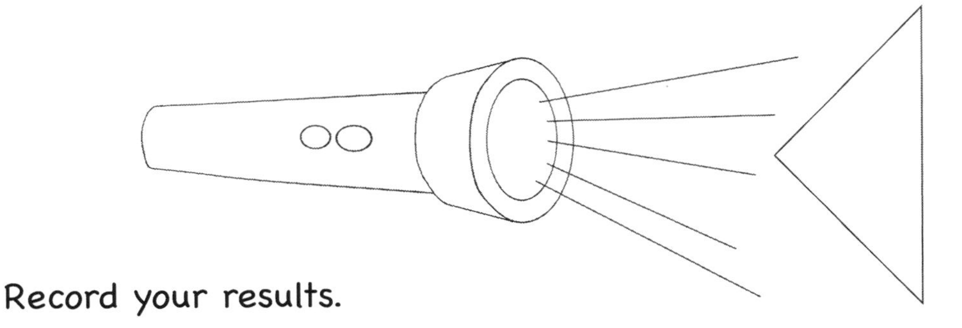

Record your results.

Observations of Light from a Flashlight Through a Prism

❷ Now take the prism and have sunlight shine through it from the same direction onto a white surface.

Record your results.

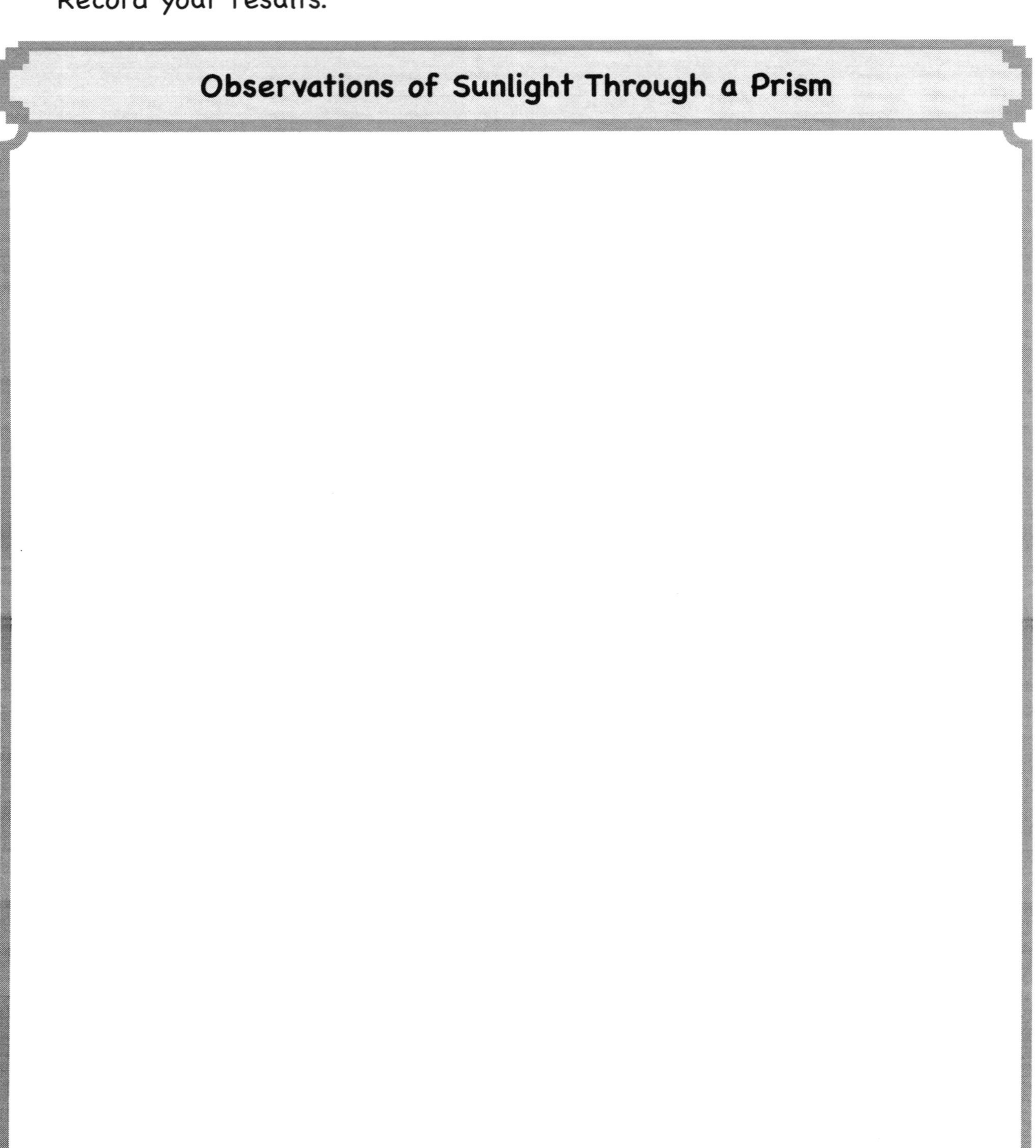

❸ Look carefully at the drawings you created for a single prism with light going through it. Knowing that red light is a long wavelength of light and that blue light is a short wavelength of light, fill in the chart to show which wavelengths are shorter and which are longer. (Hint: The longer wavelengths will be closer to red than blue and the shorter wavelengths will be closer to blue than red.)

This color of light...	has a shorter or longer wavelength than...	this color of light
green		orange
blue		yellow
orange		blue
red		violet
violet		green
yellow		red

III. What Did You Discover?

1. What happens when you shine a flashlight through a prism?

2. What happens when you shine sunlight through a prism?

3. Did you observe any differences between shining light from a flashlight and light from the Sun through a prism?

4. Do you think water droplets in the sky act like tiny prisms? Why or why not?

❺ If white light was not made up of different wavelengths and if light did not bend when it hit water droplets in the sky, would there be rainbows? Why or why not?

____

____

____

____

IV. Why?

In this experiment you explored how white light is really a mixture of different colors of light. The different colors can be separated from each other using a prism.

Sunlight contains all the different wavelengths of light and includes all the colors in the visible spectrum. As sunlight passes through a prism, each wavelength of light is bent by the prism and separated from the others. The light leaving the prism is arranged by color according to wavelength, shortest to longest—violet to red. This occurs because the different wavelengths of light travel at different speeds in transparent materials. Light waves with shorter wavelengths are traveling faster than light waves of longer wavelengths.

Because light waves travel at different speeds through a prism, they bend by different amounts depending on their wavelength and come out of the prism separated from each other. Because they are separated from each other, you can see the individual colors of the different wavelengths.

Raindrops work the same way a prism does, and when it rains you can sometimes see a rainbow of colors. The rainbow is generated by the raindrops bending the sunlight and splitting out the different colors.

V. Just For Fun

Take a second prism and place it directly behind the first one, laying it flat on one of the short edges. Using the flashlight, shine light through the two prisms together.

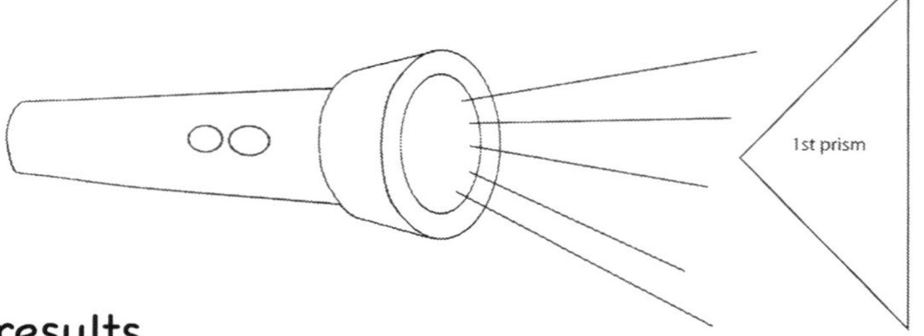

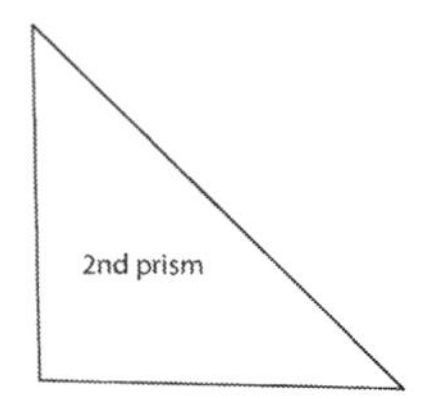

Record your results.

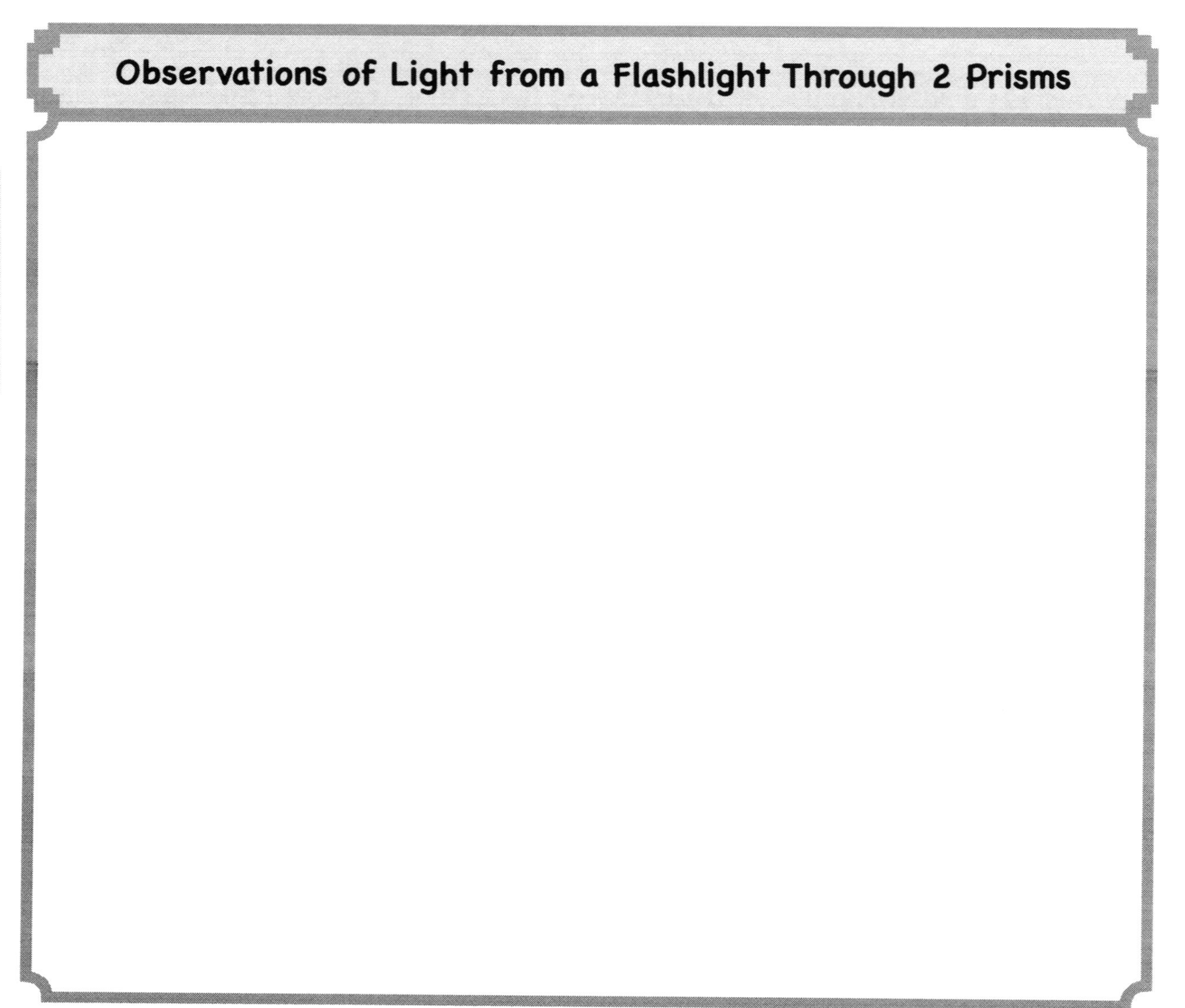

Experiment 13

Bending Light

Introduction

Try this experiment to see how light bends.

I. Think About It

1. What do you see when you look at a fishing line, stick, or pole that's partially in and partially out of the water?

____

____

____

2. What do you think the fishing line, stick, or pole will look like when you take it entirely out of the water?

____

____

____

3. Do you think if you put a spoon in your tea and the spoon appears to bend it is really bent? How can you tell?

____

____

____

4. What would happen if you put a wooden stick in vegetable oil? Would it appear to bend?

____

____

____

II. Observe It

❶ Take 3 small jars and label one **Water**, one **Oil**, and one **Acetone**. Add to the appropriate jar: 60 ml (1/4 cup) vegetable oil, 60 ml (1/4 cup) water, 60 ml (1/4 cup) acetone (nail polish remover).

❷ Add food coloring to the water.

❸ Take the glass, tilt it, and carefully pour the acetone down the side of the glass. Set the glass down on a flat surface.

❹ Put the thin wooden stick into the glass at an angle. Record your observations.

Acetone

❺ Pour the acetone back in its jar. Wash and dry the glass.

❻ Repeat Steps ❸-❺ with the vegetable oil.

Vegetable Oil

❼ Repeat Steps ❸–❺ with the water

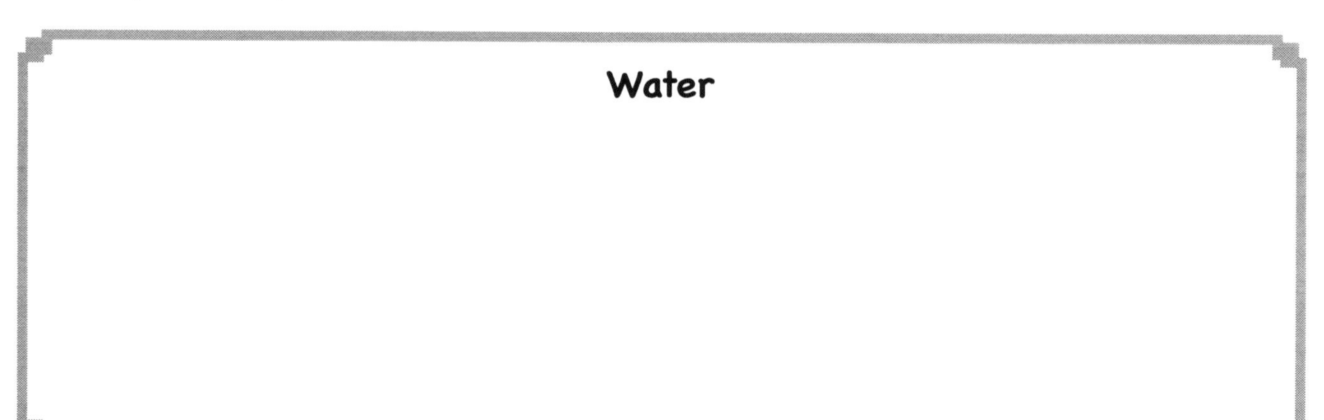

❽ Keep the glass tilted while pouring the liquids. Very slowly pour the acetone down the side of the glass. Then carefully and slowly pour the vegetable oil down the side of the glass. You should see the vegetable oil go underneath the acetone and create a nice clean layer. Carefully place the glass upright.

❾ Put the wooden stick in the glass so it goes through both layers of liquid. Tilt the stick as far as you can. Record your observations.

Acetone and Vegetable Oil

❿ Empty the glass, wash it out, and repeat Steps ❽–❾ this time using vegetable oil and water. Pour the vegetable oil in first followed by the water. Record your observations.

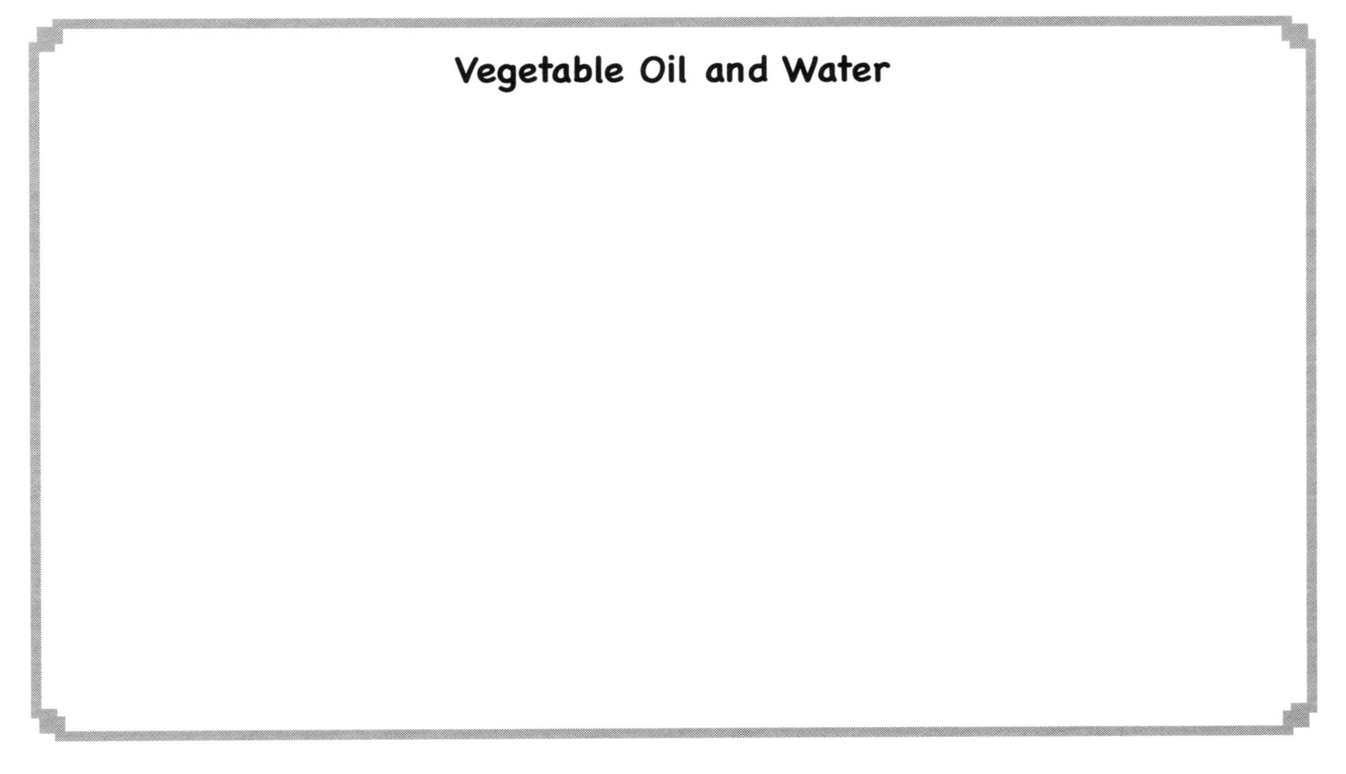

III. What Did You Discover?

❶ Did the wooden stick appear to bend in the acetone? Why or why not?

❷ Did the wooden stick appear to bend in the vegetable oil? Why or why not?

❸ Did the wooden stick appear to bend in the water? Why or why not?

❹ Did the wooden stick appear to bend again at the boundary between the acetone layer and vegetable oil layer? Why or why not?

❺ Did the wooden stick appear to bend again at the boundary between the vegetable oil layer and water layer? Why or why not?

IV. Why?

In this experiment you observed that different liquids will bend light differently. The thin wooden stick appears to bend when it enters the acetone, the vegetable oil, and the water, but the wooden stick is not actually bending, only the light coming in and going out is bending. The light is being refracted by the acetone, water, and vegetable oil. The light is also being refracted by the glass, but because you are looking at all three liquids through the same glass, what you can notice is only the difference between how the three liquids refract the light. The refraction caused by the glass stays the same.

The amount that light bends in any material is related to its change in speed. When light hits a material that slows it down, the light will bend.

It can be useful to know how fast light will travel through one material compared to how fast it will travel through a different material. Scientists use the *index of refraction* to make this comparison. The index of refraction is defined as the speed of light in a vacuum divided by the speed of light in the material. In this experiment you used liquids that have different index of refraction values *(indices of refraction)*.

The indices of refraction for the liquids you used are:

acetone: 1.36
vegetable oil: 1.47
water: 1.33

The index of refraction for a vacuum is by definition 1.00 and in air it is just slightly larger at 1.000277.

This means that light travels 1.36 times slower in acetone than in a vacuum, 1.47 times slower in vegetable oil, and 1.33 times slower in water. In this experiment, by observing how the wooden stick appears to bend differently, you can see how the speed of light varies as it travels through different liquids

V. Just For Fun

Repeat the experiment with three layers—acetone, water, and vegetable oil. This is tricky to do because you have to keep the acetone and water in separate layers or they will mix.

Keep the glass tilted while you are pouring the liquids and pour them slowly and carefully down the side of the glass. To make the layers, pour the water in first and carefully pour the vegetable oil in next. Make sure not to pour too quickly. Finally, pour the acetone in last. Again, pour slowly so you don't disturb the oil and water layers. What happens to the stick this time? Record your observations.

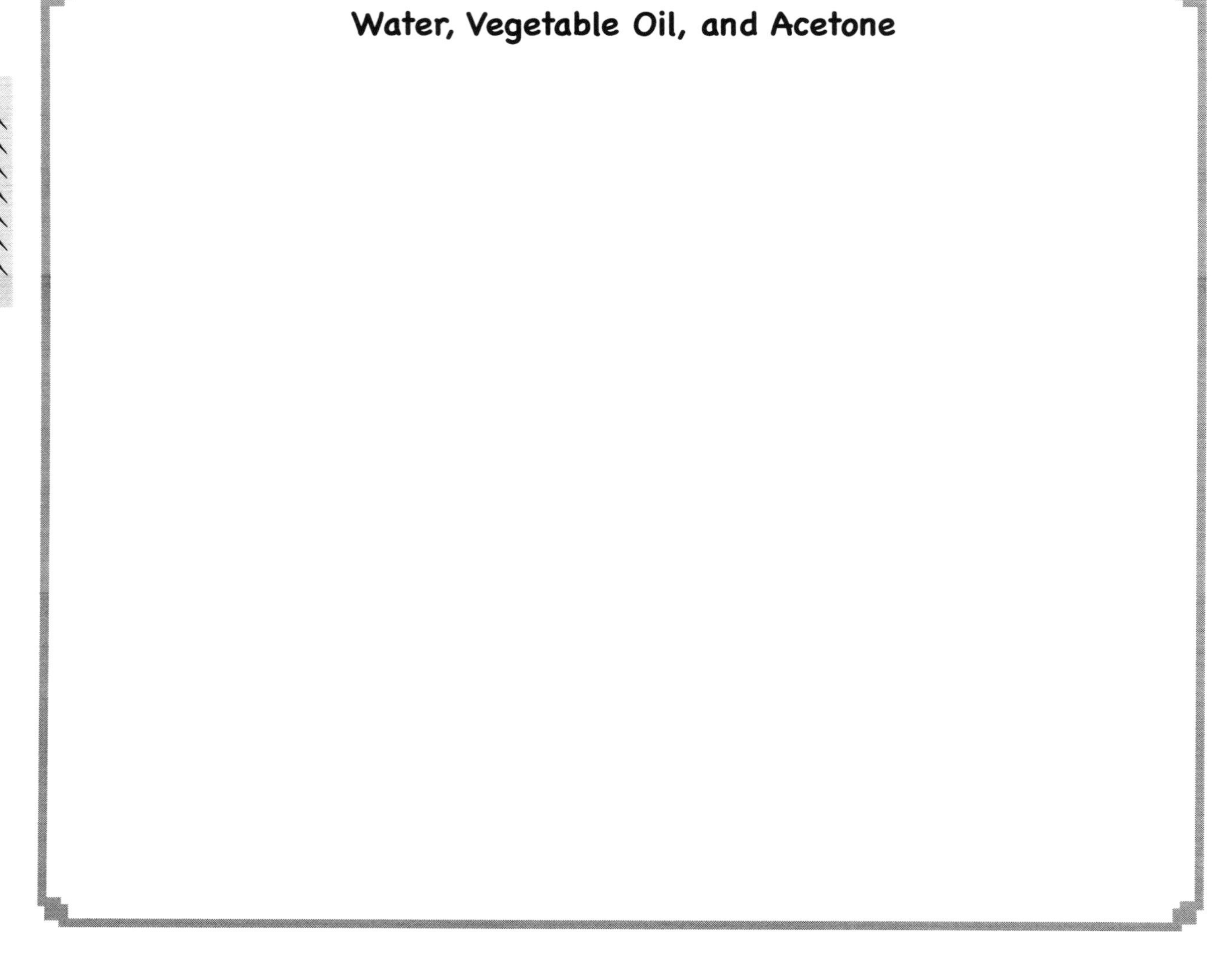

Experiment 14

Predict the Weather

Introduction

Make your own barometer—an instrument that records changes in atmospheric pressure that can be used to predict weather.

I. Think About It

1. When a storm comes in, does it get windy? Why or why not?

2. When a storm comes in, does it rain? Why or why not?

3. Describe what you have observed happening when a storm rolls in.

❹ What happens when a storm leaves?

❺ Describe the weather in your area. Does it change frequently? Is the weather different in different seasons?

❻ How can you use weather reports to help you decide what you might wear or bring with you when you go out? Can weather reports help you plan activities?

II. Observe It

1. Take a large balloon and cut the top off.

2. Stretch the top of the balloon over the rim of a glass jar and secure it with a rubber band. Make sure the balloon is stretched tightly and secured well enough that no air can get out of the jar.

3. Cut the plastic straw to about 10 centimeters (4 inches). Tape a toothpick to one end of the straw and tape the other end to the center of the stretched balloon. The end of the straw with the toothpick will extend beyond the edge of the jar.

4. Draw a line across the middle of the index card. Above the line, write "High pressure" and below the line write "Low pressure."

5. Set the jar next to a wall or a piece of propped up cardboard. Tape the index card to the cardboard or the wall, aligning the center line drawn on the index card to the position of the toothpick. The toothpick is the barometer's atmospheric pressure indicator needle.

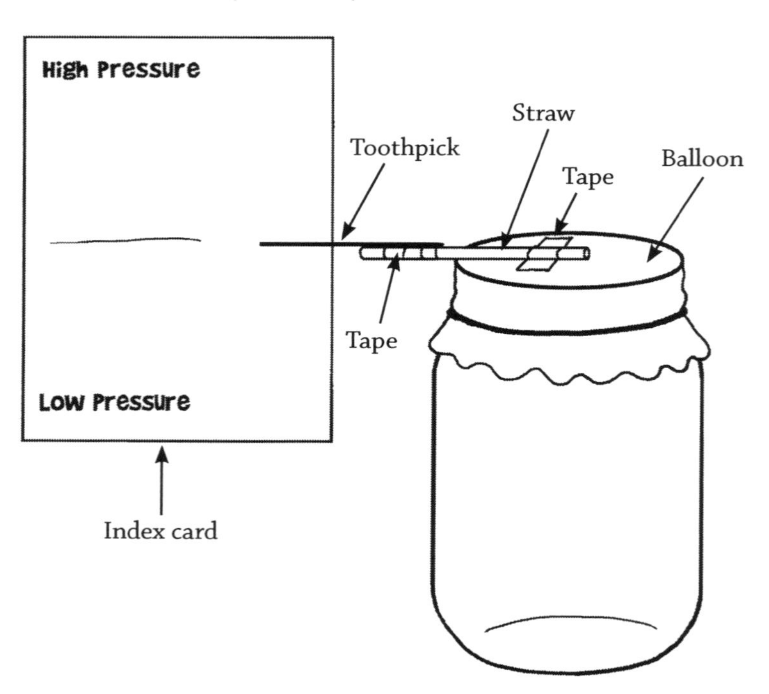

❻ Every day for a week, observe the movement of the barometer needle (toothpick) and record your observations. Record where the needle is in the "High pressure" or the "Low pressure" area of the index card, and record the weather conditions. To help you keep track of the changes in the position of the toothpick, you may want to put a mark on the index card showing the position of the toothpick and the date.

Observations—Day 1

Observations—Day 2

Observations—Day 3

Observations—Day 4

Observations—Day 5

Observations—Day 6

Observations—Day 7

❼ Review your observations from Days 1-7.

❽ Each day for the next week use the changing position of the barometer needle to predict the weather for the following day. Record the barometer reading and your predictions. In the last box note how the weather actually changed and compare it to your predictions.

Predictions—Day 1

Predictions—Day 2

Predictions—Day 3

Predictions—Day 4

Predictions—Day 5

Predictions—Day 7

Predictions compared to actual weather

III. What Did You Discover?

1. Does the toothpick move up or down on sunny days?

2. Does the toothpick move up or down on cloudy days?

3. What does the toothpick do on windy days when there are no clouds?

4. What is the position of the toothpick on hot days?

5. What does the toothpick do on cold days?

❻ Were you able to predict the weather? Why or why not?

IV. Why?

Changes in weather occur when atmospheric pressure changes. On a sunny day the atmospheric pressure is high. On cloudy days, the atmospheric pressure drops, and clouds and wind move in.

A *barometer* is an instrument that measures atmospheric pressure. The balloon barometer you made is a simple way to demonstrate changes in atmospheric pressure. Once you seal the air in the jar by stretching the balloon over it, the atmospheric pressure in the jar remains the same. Changes in the atmospheric pressure outside the jar cause the toothpick to move because the atmospheric pressure in the jar is different from the atmospheric pressure outside the jar. On a sunny day the atmospheric pressure outside the jar is higher than inside the jar, and the air outside the jar will push down on the balloon causing the toothpick to rise. On a cloudy day the atmospheric pressure outside the jar is lower than that inside the jar. This will cause the higher atmospheric pressure inside the jar to push up on the balloon, making the toothpick point down.

V. Just For Fun

Do you think clouds can be used to predict the weather? Over the next couple of weeks, observe and draw the clouds each day and note what the weather is like. When you finish your observation period, review your notes to see if you can find a relationship between the arrival of certain kinds of clouds being followed by certain kinds of weather.

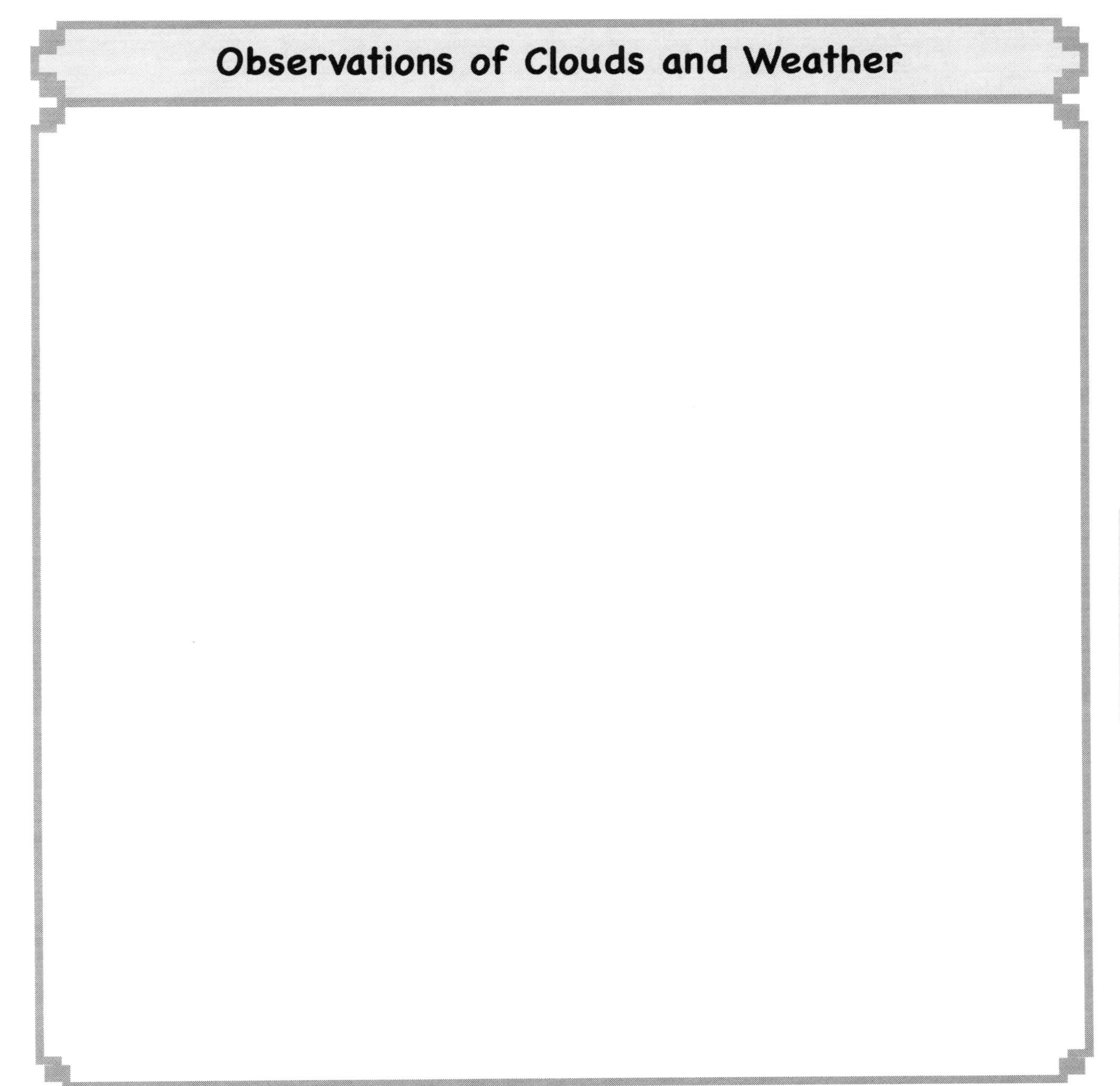

More Observations of Clouds and Weather

GEOLOGY

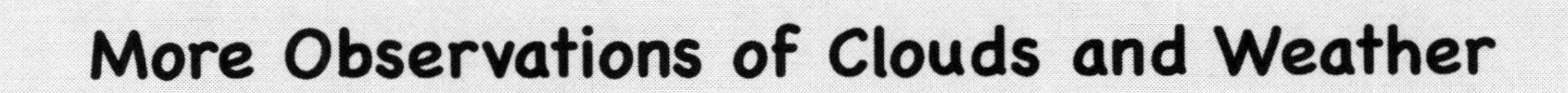
More Observations of Clouds and Weather

Experiment 15

Unintended Consequences

Introduction

Play a game about how living things in an African savanna ecosystem are interrelated. A savanna is a warm climate grassland having few trees. The African savanna is home to animals such as zebras, elephants, giraffes, various antelopes, hyenas, vultures, crocodiles, and lions.

I. Think About It

❶ What do you think would happen to lions if there were no zebras on the African savanna? Why?

____

____

____

____

❷ What do you think would happen to zebras if there were no lions on the African savanna? Why?

____

____

____

____

❸ What do you think would happen to life on the African savanna if there were a severe drought? Why?

____

____

____

____

❹ What do you think would happen to life on the African savanna if there was a very rainy year? Why?

____

____

____

____

❺ What do you think would happen if there were no scavengers on the African savanna? Why?

____

____

____

____

II. Observe It

Play the **Ruler of the Savanna** game!

❶ Cut out the **Tokens** and put them in a basket or hat.

❷ Cut out the **Unintended Consequences Cards** and put them in a different basket or hat.

❸ Cut out the two halves of the game board and tape them together.

❹ Read the **Instructions** and **Rules** and play the game.

Tokens

Cut out each **Token** along its outer edge.

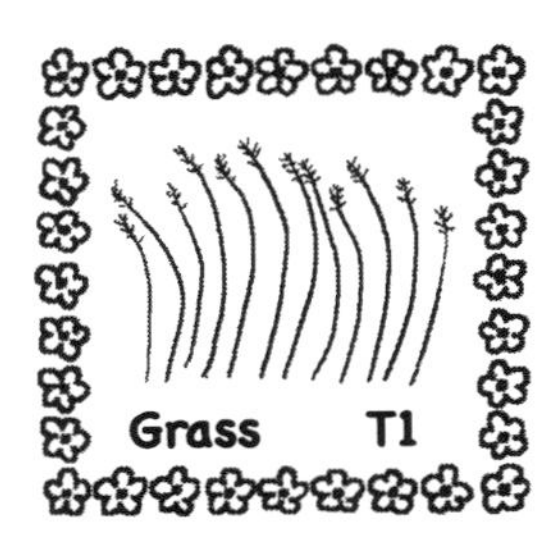

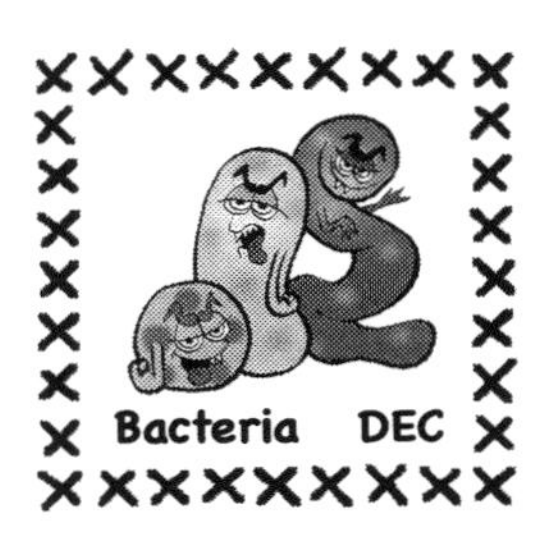

Cut out the **Unintended Consequences Cards.**

Note: The unintended consequences on these cards are hypothetical and may not be actual events.

People are concerned that the large population of grasshoppers will eat too many plants and the herbivores will not have enough to eat. The grasshoppers are sprayed with pesticides. Birds die when they eat the poisoned grasshoppers and other insects. There are more insects than ever because there are not enough birds to control them.

T3-4 take 2 steps back.

In an attempt to increase the lion population, rabbits are introduced to give the lions more food. The lions ignore the rabbits and catch their usual prey. The rabbits overpopulate the area, eating lots of grass. The zebra and other grazing populations decline from not having enough to eat, and the lions move to a new location.

T2 & T5 take 1 step back.

Cattle are introduced to the savanna. The cattle eat too much grass. Erosion results because there is not enough vegetation to hold the soil in place during rainstorms. The population of zebras and other herbivores decreases because they don't have enough food.

T2 take 1 step back

A big crate of domestic cats falls off a Land Rover driving through the savanna. The cats do very well at hunting birds and making more cats. The decreased bird population results in the insect population increasing. The population of other predators that eat birds decreases due to lack of food.

T3-4 take 1 step back.

The savanna has an unusually rainy year. All the plants and animals thrive and their populations increase.

Everyone take 2 steps forward.

Climate change makes the savanna much hotter and drier. It gets more and more difficult for plants to get enough water to live. Animals don't get enough water to drink and don't have enough plants or other animals for food.

Everyone take 2 steps back.

Hunters kill off all the lions, cheetahs, and hyenas. The zebras, antelope, and other herbivores overpopulate and eat all the grass. The rainy season comes and washes away the unprotected topsoil, creating desert conditions where little can grow.

Everyone take 2 steps back.

Red fire ants are introduced to the savanna. The ants kill plant roots while they are burrowing in the ground to make mounds. Many small animals are killed by the fire ants. Also, the fire ants take the place of the termites that were putting nutrients back in the soil.

Everyone take 1 step back.

People put out a naturally occurring fire on the savanna. The fire would have cleared out the old grass, allowing new, healthier and more abundant grass to grow. Seeds that need to be exposed to high heat before they can sprout are not able to grow. There is less food for the herbivores.

T2 take 1 step back.

UNINTENDED
CONSEQUENCES
UNINTENDED
CONSEQUENCES
UNINTENDED
CONSEQUENCES
UNINTENDED
CONSEQUENCES
UNINTENDED
CONSEQUENCES
UNINTENDED
CONSEQUENCES
UNINTENDED
CONSEQUENCES
UNINTENDED
CONSEQUENCES
UNINTENDED
CONSEQUENCES

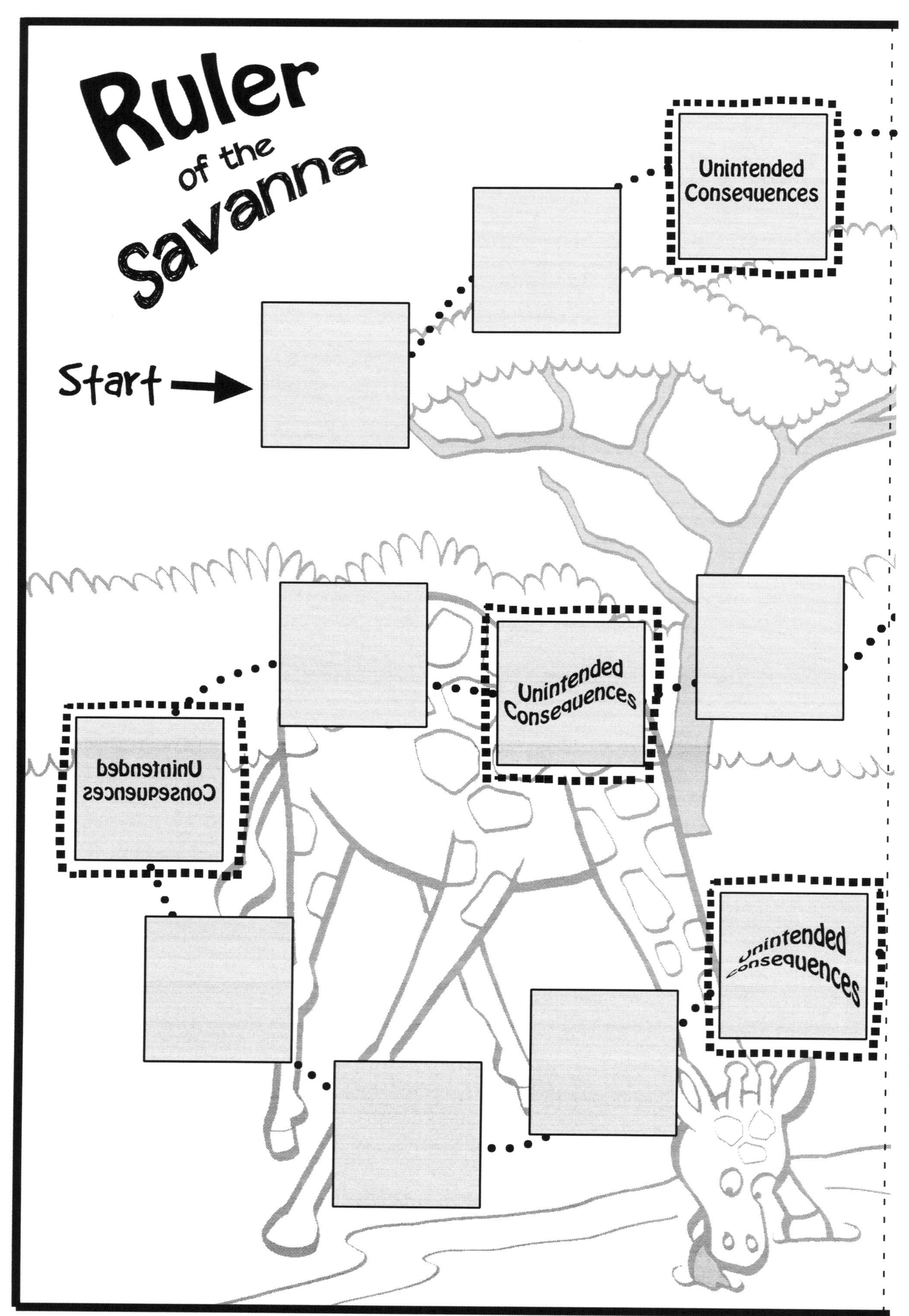
Ruler of the Savanna
Start
Unintended Consequences
Unintended Consequences
Unintended Consequences
Unintended Consequences
Cut both pages out of book. Cut this page on dotted line and tape to next page

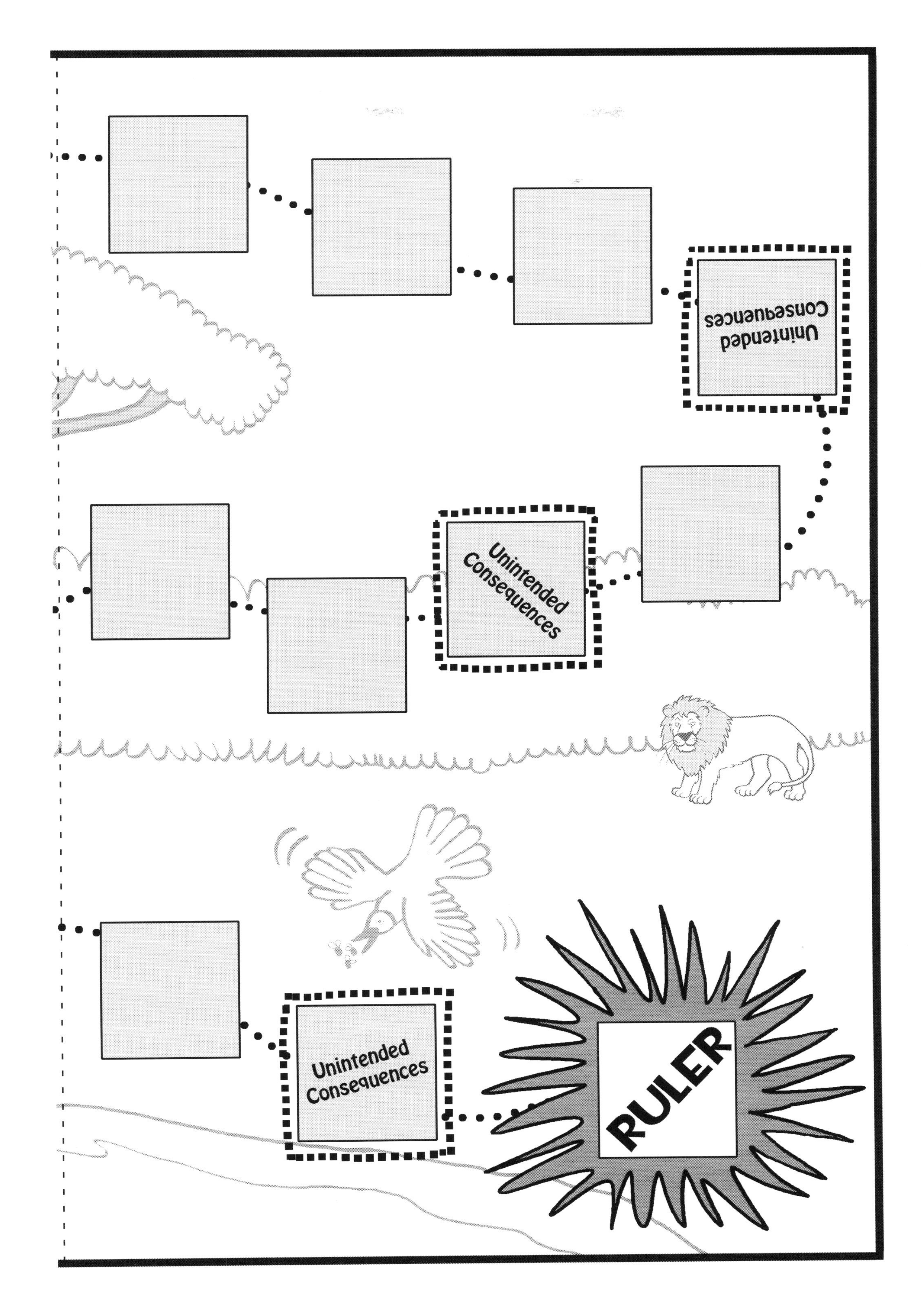

Unintended Consequences
Unintended Consequences
Unintended Consequences
RULER

Groups

The organisms on the **Tokens** are grouped by trophic level. The trophic level of each living thing is shown on its **Token** with the following symbols.

T1 Trophic Level 1 - trees, grass

T2 Trophic Level 2 - zebra, elephant, impala, giraffe

T3-4 Trophic Level 3 & 4 - hyena, vulture, cheetah, crocodile

T5 Trophic Level 5 - lion

DEC Decomposers - mushrooms, bacteria

Instructions

Each player, without looking, picks one token from the basket.

Each player throws 2 dice. Low number goes first, then the other players will follow in clockwise order.

Game is played with one die. Each player will roll the die and move their token forward that number of squares.

If a player lands on a square where there is already another token, they will look at the **Rules** to see what to do next.

If a player lands on an **Unintended Consequences** square, they will draw an **Unintended Consequences** card from the basket and follow the instructions on the card.

The first player to reach the last square wins and is **Ruler of the Savanna!**

Rules

- **T2** or **T3-4** lands on the same square as **T5**:
  T2 or **T3-4** go back 2 squares, **T5** goes forward 2 squares.

- **T2** and **T3-4** land on the same square:
  T2 goes back 1 square, **T3-4** goes forward 2 squares.

- **T1** and **T2** land on same square:
  T1 goes back 1 square and **T2** goes forward 1 square.

- Any **T** lands on the same square as **DEC:**
  T goes back 2 squares, **DEC** goes forward 1 square.

- **T3-4**, **T5**, or **DEC** land on same square as **T1**:
  nothing happens.

- Any two tokens of the same trophic level land on the same square:
  nothing happens.

III. What Did You Discover?

❶ Was there one trophic level that was more likely to win the game? Why or why not?

____

____

____

____

____

❷ Was there one trophic level that was less likely to win the game? Why or why not?

____

____

____

____

____

❸ Did the unintended consequences make it difficult to win the game? Why or why not?

____

____

____

____

____

❹ If you were to modify the game, would you add more unintended consequences? Why or why not?

____

____

____

____

____

❺ Do you think the game helped you understand how unintended consequences can affect an ecosystem? Why or why not?

____

____

____

____

____

IV. Why?

An ecosystem can be drastically disrupted by even small changes. In an attempt to improve an ecosystem, introducing an intentional change may seem like a good idea. However, it is very difficult to know all the details about how an ecosystem works because the interconnections between the animals, plants, air, physical space, and many other features are so complicated and numerous. We cannot always predict what will happen or determine the outcome of changing one part of an ecosystem. By exploring the results of changes to ecosystems that have already occurred, scientists can build models, like this game, that could help us prevent unintended consequences in the future.

V. Just For Fun

Play the game again. This time make new **Unintended Consequences Cards** using your own ideas. You can also make some new **Tokens** for different organisms of the African savanna.

—or—

Based on the **Instructions** and **Rules** for **Ruler of the Savanna**, make up your own game. You can make up a game for the ecosystem where you live or any other one of your choice. Create new **Tokens**, **Unintended Consequences Cards**, and game board.

Experiment 16

Wind or Sun?

Photo Credit: US Department of Energy, Dillon Wind Project, Iberdrola Renewables

Introduction

Build a model to explore renewable energy sources.

I. Think About It

❶ Do you have lots or little wind in your area? Are the winds always the same? Why or why not?

____

____

____

❷ Do you have lots or little sun in your area? Is the amount of sunshine always the same? Why or why not?

____

____

____

❸ Do you think you could power your home using energy from the wind? Why or why not?

____

____

____

❹ Do you think you could power your home using energy from the Sun? Why or why not?

____

____

____

II. Observe It

❶ Build your own model wind turbine or solar car! Choose one of the following kits and follow the assembly instructions.

Windmill Generator Kit: HomeScience Tools
http://www.hometrainingtools.com/windmill-generator-kit/p/KT-GSWIND/

Wind Turbine Science Kit: HomeScience Tools
http://www.hometrainingtools.com/wind-turbine-science-kit/p/KT-WINTURB/

KidWind Basic Wind Experiment Kit: Vernier
http://www.vernier.com/products/kidwind/wind-energy/kits/kw-bwx/

KidWind MINI Wind Turbine: Vernier
http://www.vernier.com/products/kidwind/wind-energy/kits/kw-mwt/

SunnySide Up Solar Car: SunWind
http://sunwindsolar.com/sunny-side-up/

Chariots of the Sun: SunWind
http://sunwindsolar.com/chariots-of-the-sun/

Solar Powered Car: HomeScience Tools
http://www.hometrainingtools.com/solar-powered-car/p/EL-SOLRCAR/

❷ Once you have built your model, test it. Note what the different parts do and how the model works. Does it do what you expected?

❸ Record your observations in the following box.

Observations of a Renewable Energy Model

Type of model ___________________________________

III. What Did You Discover?

1. How easy or difficult was it to build a wind turbine or solar powered car?

____

____

____

2. How easy or difficult do you think it is to power a full-size car with wind or solar power?

____

____

____

3. How easy or difficult do you think it is to power a home with wind or solar power?

____

____

____

4. How easy or difficult do you think it is to power a city with wind or solar power?

____

____

____

5. What problems do you think occur when trying to power a car, home or city with wind or solar power?

___

___

___

6. What other ways of generating energy do you think scientists will discover?

___

___

___

IV. Why?

By building a model wind turbine or solar powered car you can begin to understand how wind and solar energy can be used as renewable energy sources. Both solar panels and wind turbines rely on modern technologies such as photovoltaic cells and lightweight materials. Chemistry, physics, and geology provide the science behind the use of renewable energy sources, and engineering provides the design for the equipment. Chemistry, physics, geology, and engineering are all needed to develop a fully operational energy source. Semiconductors, gears, hard but lightweight plastics, and strong metals are needed to construct even small-scale solar powered cars and wind turbines.

V. Just For Fun

Pick another renewable energy kit, build it, and test it.

Compare this model to the first one you built. What are the differences? Are there similarities? Record your observations in the following box.

Can you think of any ways to modify either model that might make it work better? Try it!

Model Comparisons

Experiment 17

Take Out the Trash!

Photo credit: US Department of Energy, National Renewable Energy Lab/by David Parsons

Introduction

Explore trash!

I. Think About It

❶ How much trash do you think you make in one day?

❷ How much trash do you think you make in one week?

❸ How much trash do you think you make in one month?

❹ How much trash do you think you'll make in your lifetime?

____

____

____

____

____

❺ How many different kinds of trash do you think you make? What are they?

____

____

____

____

____

❻ Do you think you could make less trash? Why or why not?

____

____

____

____

____

II. Observe It

Find out more about the trash you make.

Week 1

❶ Keep track of all the trash you make during the week and observe what types of stuff you throw away; for example, plastic, glass, metal, and organic items like fruit peels, roots, seeds, or vegetables. Record your observations in the space provided.

Week 2

❷ For one week separate the materials you think could be recycled from those that cannot be recycled. Many items that are recyclable will have a symbol something like: ♲
Record your results.

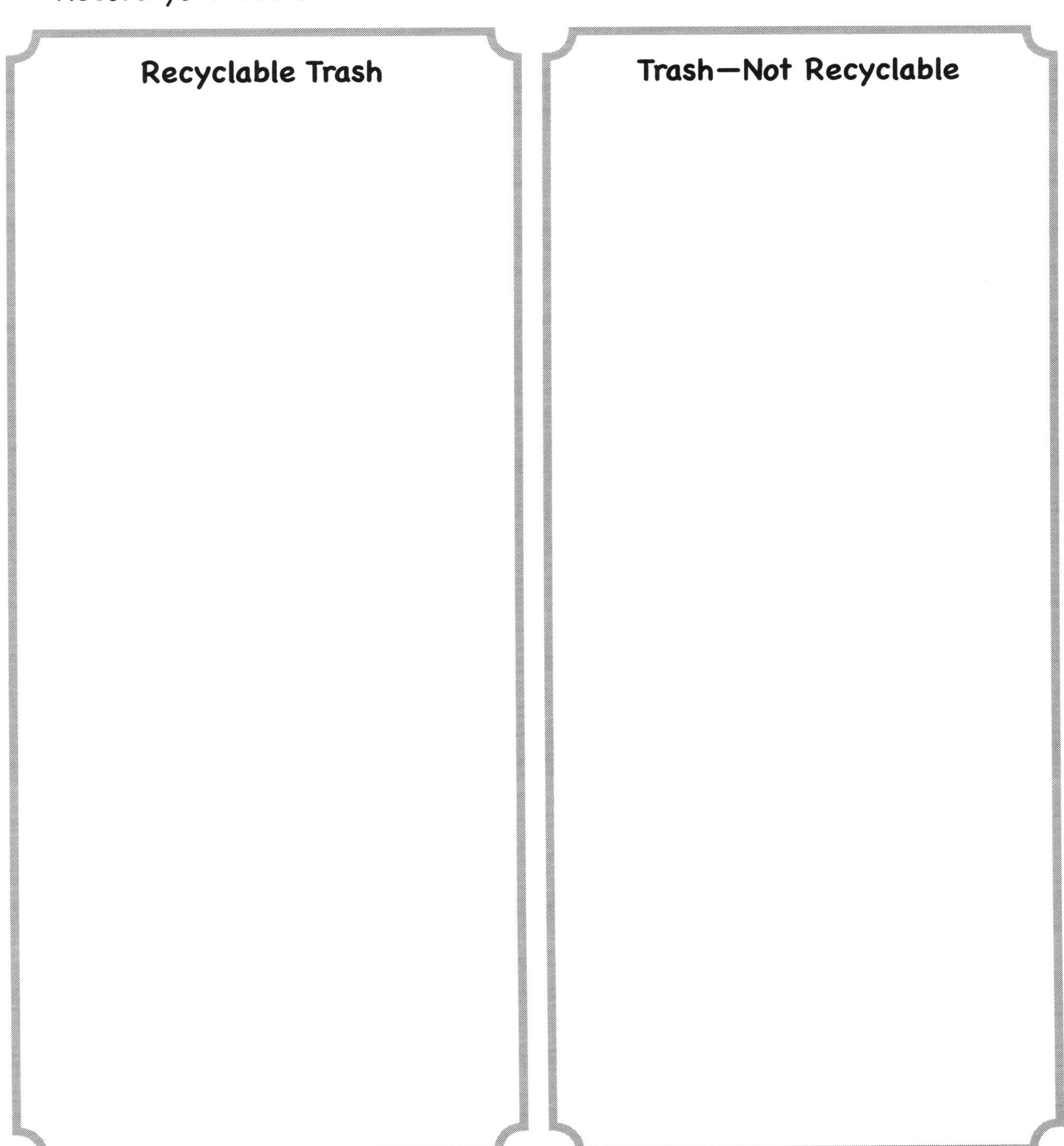

❸ Compare your trash observations for Week 1 and Week 2. Record any differences and similarities.

GEOLOGY

Week 3

❹ Keep track of all the trash you make and separate it into recyclable and not recyclable. This time as you record your observations, write down any ideas you have about how you might reduce the amount of trash you are producing.

Recyclable Trash Reduction	Trash (Not Recyclable) Reduction

III. What Did You Discover?

❶ How much trash do you create in one week?

❷ How much of your trash and what kinds of items are plastic?

❸ How much of your trash and what kinds of items are paper?

❹ How much of your trash and what kinds of items are glass or metal?

❺ How much of your trash and what kinds of items are organic?

❻ What things could you do to reduce the amount of trash you make?

____

____

____

____

IV. Why?

Many people in modern cultures don't pay much attention to the amount of trash they generate. It's easy to ignore the plastic cup, plastic bag, empty glass bottle, and paper sack that package food and other items we purchase. However, by observing what we buy and how it is packaged, we can start to change the amount of trash we produce.

To reduce the amount of trash produced, people do simple things such as bringing reusable bags with them when shopping, looking for items that are in bigger packages or have less packaging, and finding things that are in recyclable containers. They notice whether they have items they can reuse rather than throwing away or things they can take to the thrift store. They may also ask themselves if they really need to buy an item. Some people start a compost pile in their backyard so organic wastes from food, leaves, weeds, and grass mowing can turn into soil for the garden instead of going into the landfill. Some people even make art from trash!

Other solutions to the trash problem are also being investigated. For example, some companies are beginning to mine landfills to recover elements that can be reused. As well as recycling materials, this makes space in the landfill for more trash.

V. Just For Fun

Imagine that you are in charge of all the world's trash. How would you get rid of it?

Experiment 18

Up, Up, and Away!

Photo credit: NASA

Introduction

Experiment with rocket building!

I. Think About It

❶ How would you design a homemade rocket?

____

____

____

❷ How high do you think a homemade rocket could go? Why?

____

____

____

❸ What type of fuel would you use to make a rocket? Why?

____

____

____

❹ Do you think a homemade rocket could reach the Moon? Why or why not?

____

____

____

II. Observe It

Choose one or more of the suggested model rocket kits and follow the instructions to build and test a model rocket.

High-Fly Rocket Kit: HomeScience Tools
http://www.hometrainingtools.com/high-fly-rocket-kit/p/KT-FHROCK/

001225 - **Alpha Rocket**: Estes Rockets (engine sold separately)
http://www.estesrockets.com/rockets/001225-alphar

000651 - **Der Red Max Rocket**: Estes Rockets (engine sold separately)
http://www.estesrockets.com/rockets/kits/skill-1/der-red-maxtm

000810- 220 **Swift Rocket**: Estes Rockets (additional parts sold separately)
http://www.estesrockets.com/000810-220-swifttm

III. What Did You Discover?

❶ How easy or difficult was it to build a model rocket?

____

____

____

____

❷ What problems did you encounter building the rocket?

____

____

____

____

❸ How high did your rocket fly?

❹ How do you think you could get your rocket to fly higher?

❺ What do you think would happen if you increased the payload?

❻ What would you have to do to get your rocket to fly to the Moon?

IV. Why?

Model rocket are a great way to gain more understanding about how full-size rockets work. By observing how the model works when you test it, you can learn more about different parts of a rocket and how they function, such as why the body of the rocket needs to be a particular shape, the need for fins, and how the burning of fuel launches the rocket into the air.

A rocket can travel from a few miles to several hundred miles above Earth and even out into space, depending on the design of the rocket, the fuel used, and the payload carried. Many factors must be taken into account when designing a rocket. It needs to follow a predetermined path and also must respond to control from the ground. The rocket has to contain the proper amount and types of fuel to launch it to the required altitude. It needs to have a powerful enough propulsion system to lift a particular payload that may consist of equipment, supplies, and astronauts in a command module.

A command module is the part of a spacecraft where the astronauts live, control the flight, and communicate with the ground. A command module has its own rockets to propel it and to change the direction in which it is moving. It must be carefully designed in order to support life and return astronauts safely to Earth.

Many rockets are unmanned. These may carry satellites that will be released to orbit Earth, they may bring supplies to a space station, or they may carry landers to Mars.

V. Just For Fun

Think of changes you could make to redesign your rocket. Could you add to the payload, change the frame or the fuel? What else could you change? Make one change at a time, test the rocket after each change, and observe what happens.

Record your ideas and results.

Ideas for Redesigning a Rocket

Results of Redesigning a Rocket

Experiment 19

How Far?

Introduction

Learn how to use the parallax effect to determine how far away a distant object, trail, or road is from where you are standing.

I. Think About It

❶ As you look across the room, how far is it to the table, desk, or chair?

____

____

____

❷ Look across the street. How far is it to the building, tree, or car parked on the other side?

____

____

____

❸ Look at a distant tree, building, or other object. How far away is it from where you stand?

____

____

____

❹ Do you think you can use parallax to estimate how far away something is? If so, how do you think you could do it?

____

____

____

II. Observe It

❶ Practice observing the parallax effect by using your thumb. Pick an object some distance away. Hold your arm at full length in front of you with your thumb up. Close one eye.

❷ Without moving your thumb, switch eyes, closing the one that is open and opening the one that is closed. Record your observations below and note if your thumb appeared to move. If it did, note how much it appeared to move.

❸ Find the ruler at the end of this *Observe It* section, cut it out, and tape it together with "0" in the center of the ruler.

❹ Pin or tape the ruler to a wall.

❺ In this experiment you will use your feet as measuring devices. Each of your feet from heel to toe is one "foot." Starting with your back against the wall, walk heel to toe 5 "feet" away from the wall and then turn around.

❻ Hold your arm at full length, thumb up. Close one eye and align your thumb with the "0" marked in the middle of the ruler.

❼ Switch eyes and observe the position on the ruler that your thumb now lines up with. In the table below, record how far your thumb appears to have moved from "0."

❽ Walk another 5 feet (for a total of 10 feet), and repeat Steps ❻–❼.

❾ Walk another 10 feet (for a total of 20 feet) and repeat Steps ❻–❼.

Number of "feet" from wall	Distance of thumb from "0"
5 feet	
10 feet	
20 feet	

⑩ Take the three measurements you recorded in the table and graph your data on the following chart.

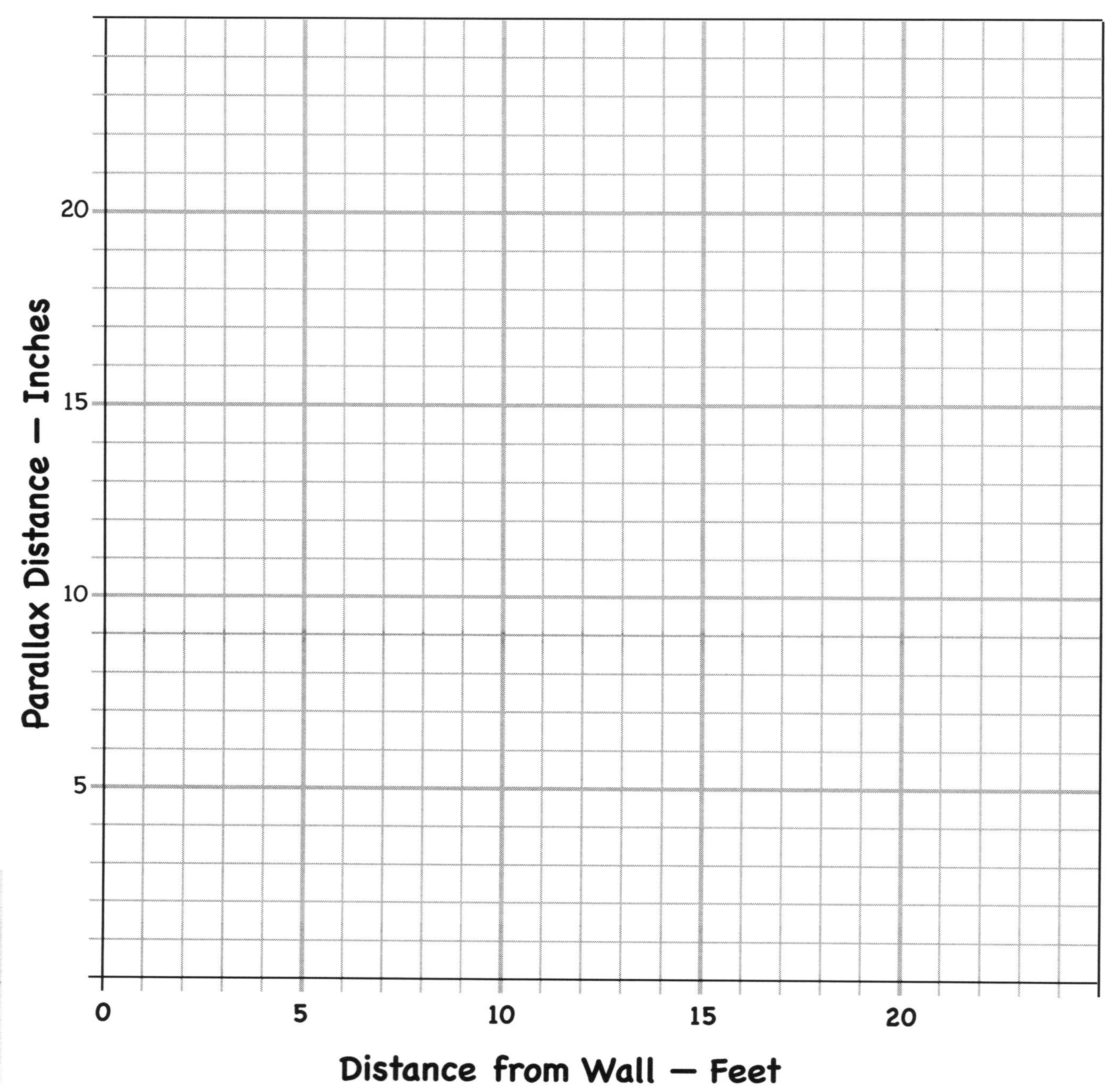

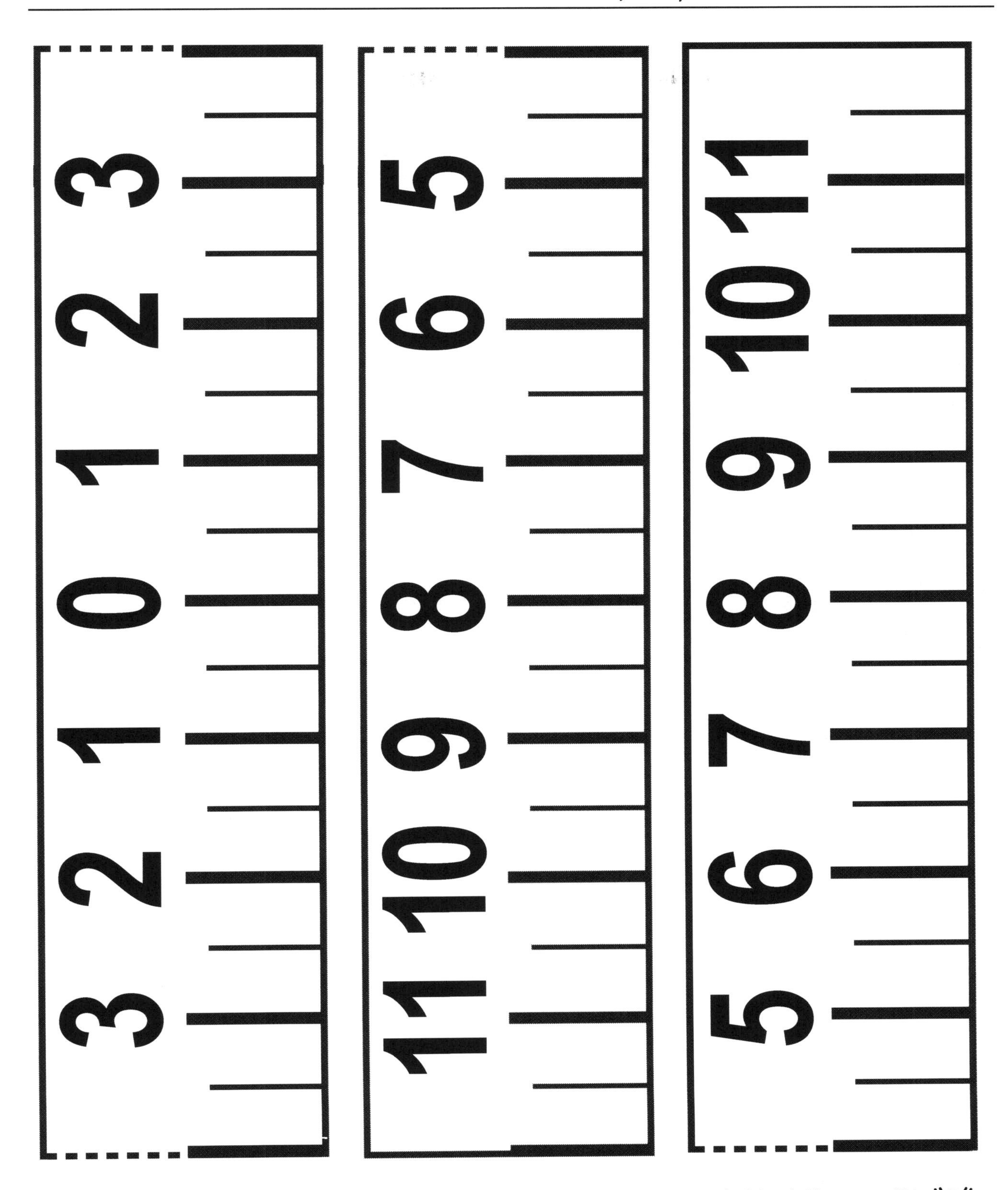

Cut out the ruler sections and tape together at the dotted lines with "0" in the center of the ruler. Fill in the missing number 4 on each side of the ruler.

III. What Did You Discover?

1. How many inches did your thumb move when you were 5 feet away from the wall?

   ____

2. How many inches did your thumb move when you were 10 feet away from the wall?

   ____

3. How many inches did your thumb move when you were 20 feet away from the wall?

   ____

4. Did the parallax distance of your thumb increase or decrease as you walked farther away from the wall? Why?

   ____
   ____
   ____
   ____

5. On the graph, can you draw a straight line through the three points you plotted? Why or why not?

   ____
   ____
   ____
   ____

IV. Why?

In this experiment you observed that the degree of parallax is proportional to the distance. In other words, as the distance between you and the object you were observing (the ruler on the wall) became greater, the amount your thumb appeared to move as a result of the parallax effect also became greater.

Each time you extend your arm fully, your thumb stays the same distance away from your face. As you peer at your thumb, you can make a good guess how far away something is by observing how much your thumb appears to move when you switch eyes. In fact, the length of your arm is about 10 times the distance between your eyes, and you can estimate how far away something is by multiplying by 10 the distance your thumb appears to move.

For example, if you are looking at a gate that is 5 feet (1.5 meters) wide and when you switch eyes, your thumb appears to move about the width of the gate, you can multiply this width times 10 and estimate that you are 50 feet (15 meters) from the gate. Or, if you are looking at a barn that is 100 feet wide and the distance your thumb moves is about 2 barn widths, you can estimate that you are 2000 feet from the barn (100 ft. x 2 x 10). (In meters, 30 m x 2 x 10 = 600 meters away)

Using the parallax effect with your thumb is a great way to estimate distances to objects you can see on Earth. Although the distances you calculate won't be exact, they will be close enough to help you decide if you need to sit and have a snack before you tackle that next mountain peak!

V. Just For Fun

Using your thumb, try to determine the distance of an object. Pick a large object such as a building, wall, or car. Guess how wide the object is and then use the method described in the *Why* section to estimate how far away the object is.

You can do this using feet or meters, but if you don't know the size of the object, you can use the object itself as a measurement. For example, if you are looking at a car and you can fit the length of 3 cars between the two positions of your thumb, you can calculate that the car is 30 car lengths away from you.

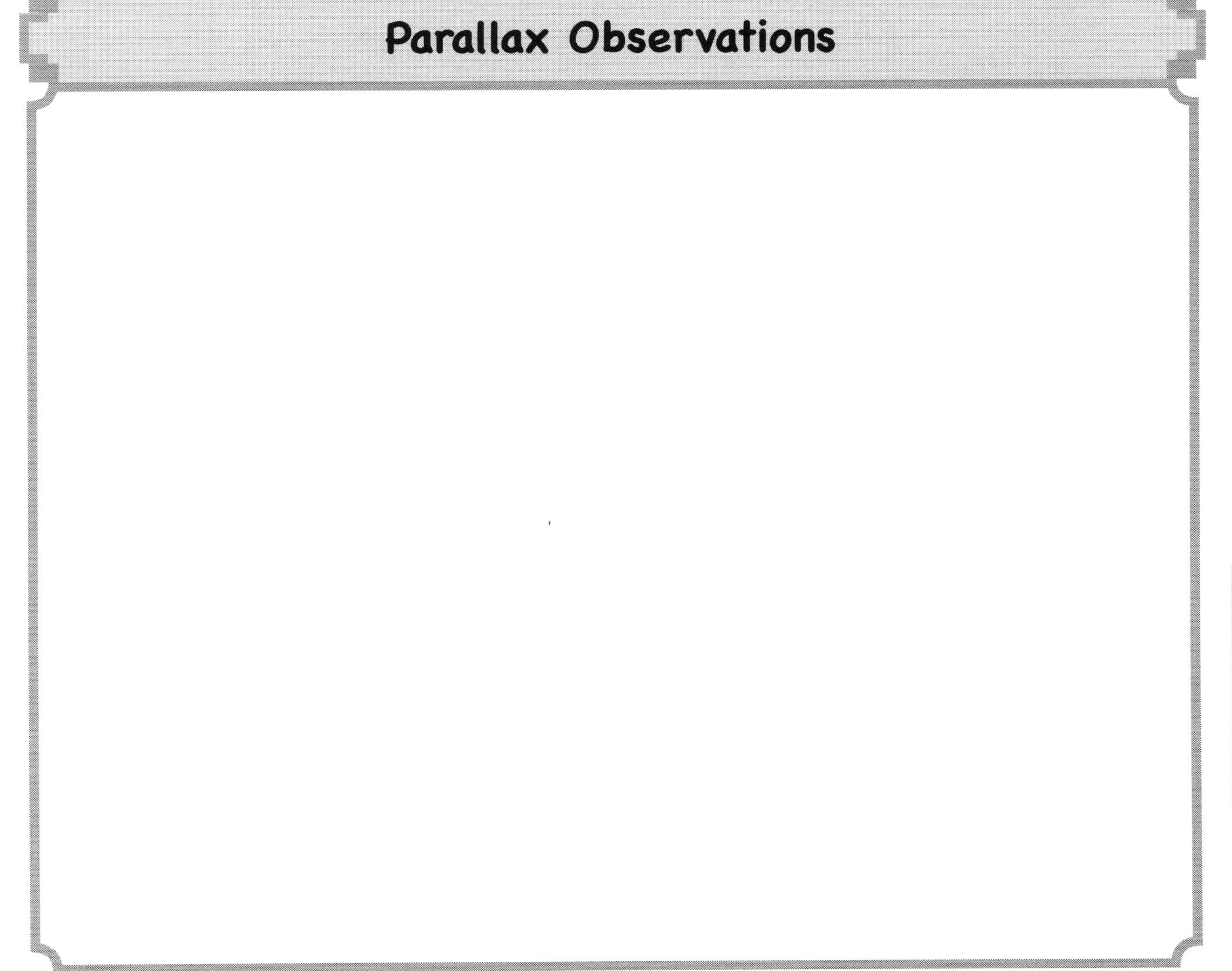

Experiment 20

Finding Exoplanets

Kepler Spacecraft

Artist's illustration courtesy of NASA/Ames/JPL-Caltech

Introduction

Help astronomers discover exoplanets!

I. Think About It

❶ What tools do you think scientists could use to find planets that are outside our solar system?

____

____

____

❷ What advances in technology do you think would make detection of exoplanets easier?

____

____

____

❸ Do you think there are exoplanets that could support life? Why or why not?

____

____

____

❹ How would you search for intelligent life on exoplanets?

____

____

____

II. Observe It

Exoplanet Explorers is an astronomy project on the Zooniverse website. This project allows everyone to become a scientist. Use data collected from the Kepler spacecraft K2 mission to help astronomers find exoplanets.

1. Go to the Zooniverse website at www.zooniverse.org.
2. Click on the **Register** tab on the top menu bar and fill in the required information to become a Zooniverse citizen scientist. In the box on the next page, write down your login name and password so you can sign on to Zooniverse again later.
3. Once you have registered, click on the **Projects** tab on the top menu bar, then click the right arrow to find the **Space** icon. Click on this icon and then on the **Exoplanet Explorers** K2 project.
4. Read the **About** pages, including the **FAQs**. Explore the **Talk** and **Collect** areas.
5. Once you have explored the different project areas, click on the **Classify** tab. A box titled **Welcome to Exoplanet Explorers** will appear. This is the project tutorial. If you don't see this box, click the **Show Project Tutorial** box at the bottom of the screen.
6. Read all the pages of the tutorial.
7. Click on the **Field Guide** tab on the right to help you become familiar with the characteristics you are looking for. Clicking on each example type will bring up the information for it. (Use the back arrow at the top to return to the example choices.)
8. Start classifying transiting planets. Look at the images on the left side of the screen and select **yes** or **no**, then click **Done**. If you want to change your mind, click **More**, otherwise click **Next**.
9. Use the box on the next page to make notes and sketches of your observations.

Login Info, Notes, and Sketches

III. What Did You Discover?

❶ Was it easy or difficult to classify potential exoplanets? Why?

❷ Did you find analyzing the different data files interesting, tedious, or both? Why?

❸ Do you think you were you able to identify a transiting planet? Why or why not?

❹ What did you learn from the **Field Guide**?

❺ How much time did you spend looking for exoplanets? Why?

IV. Why?

New technologies, such as the Kepler spacecraft, are gathering vast quantities of data. Due to the limitations of computers, some of this data must be analyzed by actual humans. It is now possible for members of the general public to play an important role in scientific discovery by becoming citizen scientists. Having hundreds or thousands of people participating in a scientific project means that a larger amount of data can be analyzed in a far shorter period of time than if only a few people were performing this task. By participating in citizen science, you could be part of the next great scientific discovery!

This experiment uses data collected from the Kepler spacecraft's K2 mission. The Kepler spacecraft collects data about the brightness of a star. Changes in the star's brightness can be used to find exoplanets by the transit method. Recall that a transit happens when a planet crossing in front of its sun produces a small change in the star's brightness. These small changes can be detected and graphed.

It can be challenging to sort through millions of data files to find a transiting planet. The data files are made up of a collection of points that may or may not fit the pattern for a transiting planet. It can be tricky to decide which data files fit the pattern for a transiting planet and which do not. Also, there are far more data files that do not fit the pattern than there are data files that do. Looking through these files can become tedious and frustrating.

However, real scientific discoveries often happen from being patient and looking through volumes and volumes of scientific data. The more you look at the exoplanet data files, the more familiar you become with what the information in the files represents. After spending time analyzing the data files, you become better and better at quickly determining what may or may not fit the profile of a transiting planet.

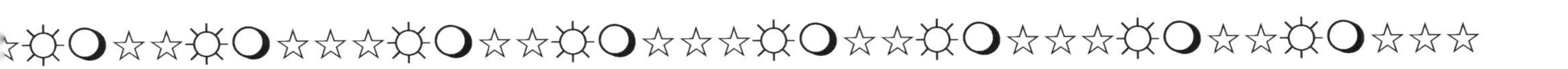

V. Just For Fun

Imagine you are an astronomer looking for signs of intelligent life on exoplanets and you can use any kind of technology you can think of, including your own new inventions. How many different ways could you look for signs of life? What new equipment could you invent? Could you combine different kinds of technology to gather more information? If you found intelligent life, how would you try to communicate with it? How would it try to communicate with you?

Record your ideas and drawings. There's more space on the next page.

How to Look for Life on Exoplanets

Experiment 21

Build a Robot!

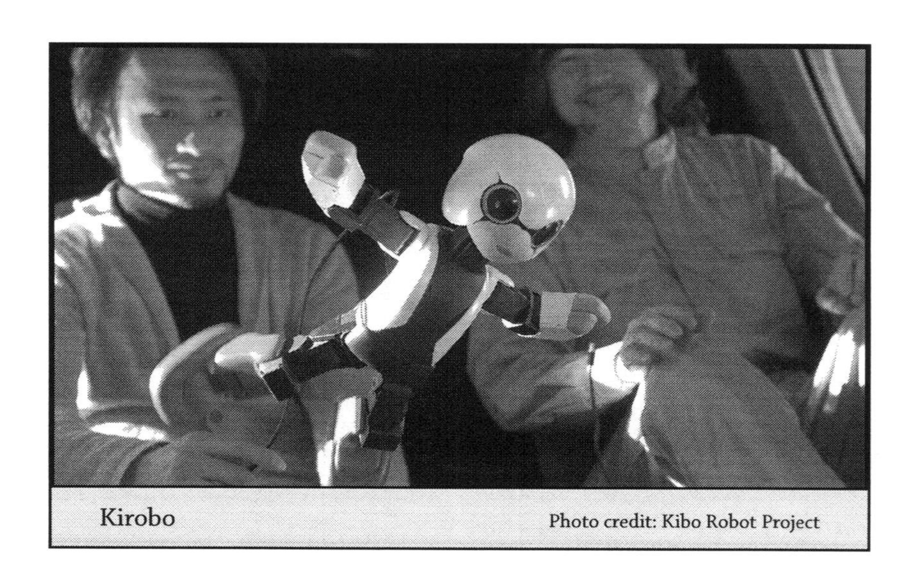

Kirobo   Photo credit: Kibo Robot Project

Introduction

Learn more about robots by building one!

I. Think About It

❶ Do you think it is easy to build a robot? Why or why not?

____

____

____

❷ What features do you think a robot needs? (for example, the ability to move forward and backward, etc.)

____

____

____

❸ Is there anything special you would need to consider if you were building a robot that could go into space?

____

____

____

❹ If you were to design a robot for space what would it look like? What would it be able to do?

____

____

____

II. Observe It

❶ Assemble a robot using any of the following robotics kits or another of your choice

http://www.hometrainingtools.com/robo-link-robot-kit/p/KT-ROBLINK/

http://www.hometrainingtools.com/jungle-robot-kit/p/EL-JUNGLE/

http://www.robotshop.com/en/2wd-beginner-robot-chassis.html

http://www.robotshop.com/en/em4-educational-motorized-robot-kit.html

❷ Assemble your robot, then observe what it can do (for example, walk, run, jump, etc.). Record your observations.

❸ Observe what your robot cannot do (for example, walk, run, jump, etc.). Record your observations below.

❹ Note how your robot is powered (for example, with a battery, solar, hydroelectric, nuclear energy, etc.). Record your observations.

❺ Record the types of materials your robot is made of (for example, plastic, metal, glass, etc).

❻ If you could modify your robot, what changes would you make to it? What other things would you have it do? Record your ideas.

III. What Did You Discover?

❶ How easy or difficult was it to assemble your robot? Why?

____

____

____

❷ What things can your robot do?

____

____

____

❸ What is your robot not able to do?

____

____

____

❹ How long does the power for your robot last?

____

____

____

❺ Would you be able to refuel your robot if it were in space? Why or why not?

____

____

____

IV. Why?

Using robots in space is an exciting option for exploration. However, designing robots that can actually carry out sophisticated functions can be challenging. In addition to designing robotic functions, scientists and engineers need to be able to build systems that allow them to operate the robots from great distances.

By building your own robot you can experience some of the challenges scientists and engineers face while designing robots for space exploration. It is easy enough to use gears and wheels to make a robot that rolls. But what about a robot with appendages, like arms and legs, that bend so it can climb stairs or pick up objects? What about making a robot that can see both close and far distances, like the human eye, or a robot with delicate hand movements that can operate tools and equipment? What about robots that can recognize people's faces and learn languages? What about the possibility of a robot that can solve problems by thinking about them? By asking these kinds of questions, scientists and engineers find new areas of robotics to explore.

V. Just For Fun

If your robot walks or rolls, try to attach a "payload" to the robot. Can it carry a pencil? Drag a rock? Push a book? What else can it do with a payload?

Can you modify your robot to perform any additional functions? Try it!

Record your observations in the following space.

Observations of a Robot

Experiment 22

Winning the Nobel Prize

Introduction

Think about the Nobel Prize while reviewing your experiments.

I. Think About It

❶ What do you think it would be like to win the Nobel Prize?

____

____

____

❷ What would you say during your Nobel Prize speech?

____

____

____

❸ In what field of study would you want to win the Nobel prize? Why?

____

____

____

❹ How do you think the Nobel Prize helps keep scientists working towards discoveries that are of the "greatest benefit to mankind?"

____

____

____

II. Observe It

❶ Take a look at all of the experiments you did for this book. Which experiments worked? Record your observations below.

Experiments that Worked

❷ Take a look at all of the experiments you did for this book. Which experiments did not work? Record your observations.

Experiments that Did Not Work

❸ Look at your lists of experiments that worked and those that didn't work. What new discoveries did you make by doing the experiments? Record your observations.

Discoveries

❹ Pick one experiment you did that did not work. Discuss why it did not work and what you might do to make it work better.

Discussion of an Experiment that Did Not Work

❺ Look at all of your experiments and find those that involved multiple areas of science. Were there experiments that used both chemistry and physics? Chemistry and biology? Geology and astronomy? Other combinations? Record your observations below.

Experiments Using Multiple Areas of Science

III. What Did You Discover?

1. How do you define whether an experiment worked?

____

____

____

____

____

2. How do you define whether an experiment did not work?

____

____

____

____

____

3. How do you define a discovery?

____

____

____

____

____

❹ Do you think you would make more discoveries by playing more with the experiments, whether they worked or not? Why or why not?

❺ How might you use what you've learned in this book to make an important discovery?

❻ How do you think it will help you make important discoveries when you know how all the areas of science connect with each other?

IV. Why?

Making discoveries is the most exciting part of doing science. Discoveries in science start out as small observations that bring up new problems to be solved and new questions to be answered. Many scientists focus on one area of science, but knowing that many areas of science connect to each other creates more opportunities to make important discoveries. For example, Marie Curie was awarded both a Nobel Prize in physics and a Nobel Prize in chemistry for her work with radium and polonium that involved both chemistry and physics.

Important discoveries always begin with good questions. But these questions don't have to be complicated. Many times they are very simple. Albert Einstein asked the question, "What would it be like to ride a rainbow?" By asking this simple questions and then doing thought experiments about it, Einstein discovered new ways to think about the nature of light and the way light travels in space.

Nobel Prize winning projects start with questions that aren't always complicated and sometimes the discoveries can be small, just like the discoveries you made using this book!

V. Just For Fun

Imagine that you are being considered for the Nobel prize. You are one of the few scientists to receive a Nobel prize nomination letter. Your paperwork has been submitted, and you are waiting for the results from the Nobel Committee.

Pick your favorite experiment. Plan a presentation to the Nobel Committee telling them about what you discovered by doing this experiment and why you think it's important. Think about how your discoveries might be used. Write, draw, or make a video of your presentation.

Use the following space to record your ideas.

My Nobel Prize Presentation